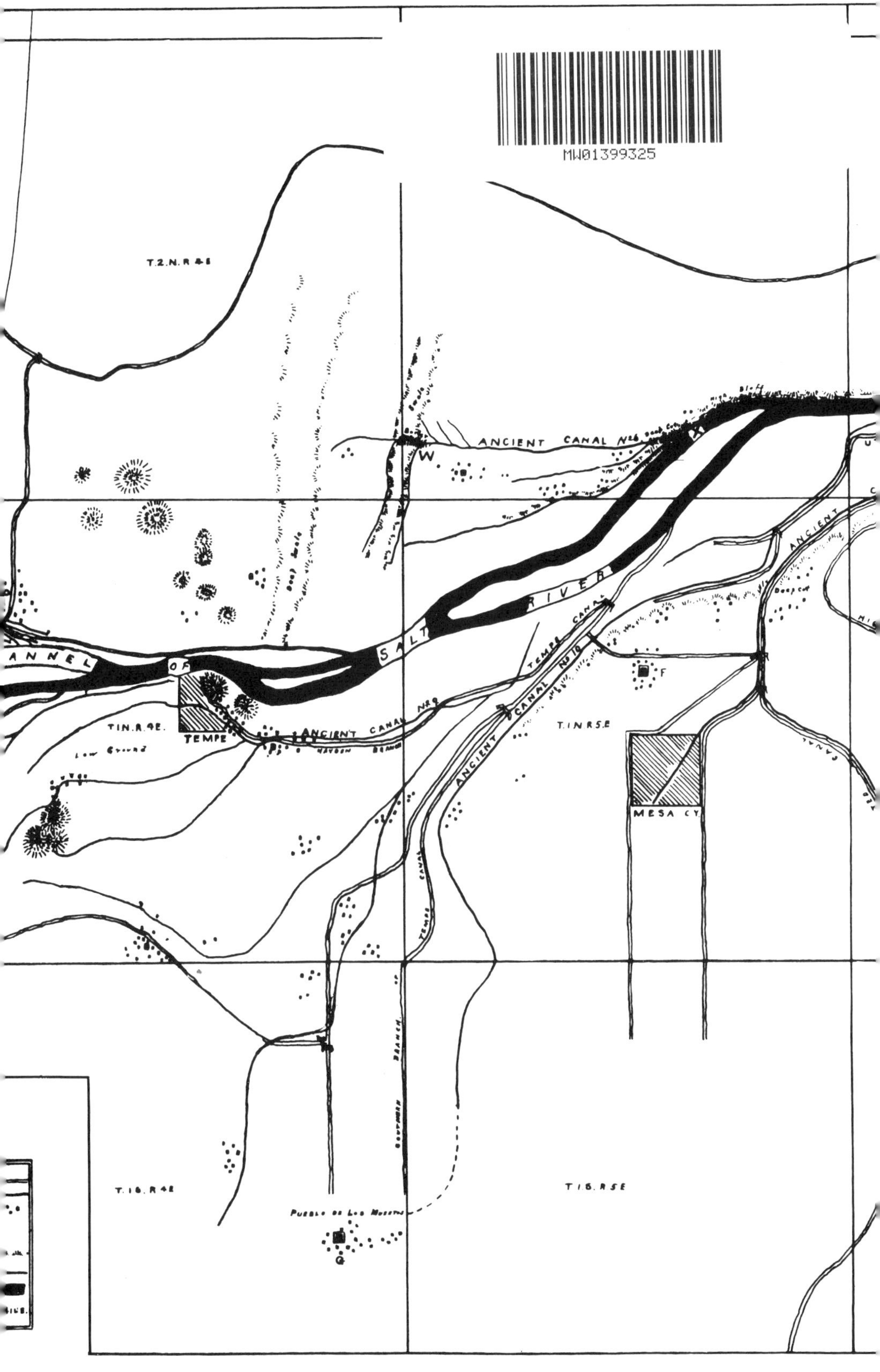
T.2.N.R.4.E
ANCIENT CANAL No 6
W
X
ANNEL
OF
SALT
RIVER
TEMPE CANAL
ANCIENT CANAL No 10
F
ANCIENT
TEMPE
T.1.N.R.4E.
ANCIENT CANAL No 9
HAYDEN BRANCH
Low Ground
T.1.N.R.5.E
MESA CY.
ANCIENT CANAL
TEMPE CANAL
SOUTHERN BRANCH OF TEMPE CANAL
T.1.S.R.4E
T.1.S.R.5E
Pueblo de Los Muertos
G

New Insights
to
Antiquity

NEW INSIGHTS TO ANTIQUITY

A DRAWING ASIDE

OF THE VEIL

WITH

58 PLATES AND 10 FIGURES

BY

RICHARD PETERSEN

ENGWALD & CO. • 4730 E. INDIAN SCHOOL RD., STE. 120
PHOENIX, ARIZONA • 85018

Copyright © 1998 by Richard G. Petersen
All rights reserved under international
and Pan-American copyright conventions.

Published in the United States by
Engwald & Co.
4730 E. Indian School Rd. Ste. 120
Phoenix, Arizona   85018

ISBN 0-9662134-1-6
Library of Congress Card No. 97-77780

KEY WORDS: Catastrophism, Comets, Atlantis, Ice Ages

Manufactured in the United States of America

CONTENTS

	Preface	vii
	Acknowledgements	ix
Prologue:	A Brief Introduction	1
Chapter 1:	The Great Cities	7
2:	Expedition of Conquest	31
3:	The Sequel	49
4:	Dead Men	73
5:	Signs of Catastrophe	97
6:	Ice-age Residues?	119
7:	Opening the Door	149
8:	The Great Destroyer	173
9:	A Fateful Rendezvous	193
10:	Easter Island	211
11:	Atlantis, et Cetera	229
12:	The Great Flood	247
Epilogue:	Afterthoughts	263
Appendix A:	The Tides	281
B:	A Big Bang?	293
C:	The Geomagnetic Field	301
D:	Volcanism	309
E:	In Re. Big Bones	317
	References	327
	Index	333

PREFACE

ALTHOUGH THIS BOOK deals most directly with riddles from the ancient past it relates by implication to some of the most pressing social problems that confront us today. In fact, these two sets of problems are so intimately connected that the latter cannot be resolved without first resolving the former. I will not attempt to justify this assertion here because I anticipate that in due course the reader will reach this conclusion for himself without any help from me. I mention the point only to explain the significance that I place upon these findings and the burden that I feel to present them with the greatest possible integrity.

Because these findings should be of concern to the public at large I have arranged the discussion so it will be accessible to readers having no particular technical background. To this end I have set aside as Appendices at the end most of those segments that do call for some formal preparation. Consequently, despite the fact that a great many topics enter into this discussion, the main body of the presentation should be accessible to almost any reader.

The central thrust of this work is to expose and correct certain serious errors about earth's past now current in the ivied halls of learning. Grievous errors they are indeed, but like old shoes one becomes accustomed to them. Where they chafe he grows callouses and becomes numb to the problems. This very fact, that modern Science at its sophisticated best is blind to long-standing error, will come as a shock and an embarrassment to many. I am keenly aware that no one likes to see his mistakes aired in public so I can hardly expect these findings to be gladly received by everyone. Dignities and reputations are at stake—but so also is the general welfare, as noted above. Balancing one interest against the other I conclude that the common good is best served by allowing the truth to prevail; nevertheless, I do regret any embarrassment that may result from these revelations.

Finally, a few words about the subtitle of this book might be in order. It is a rendering of the Greek word, *Anacalypsis*, used by Godfrey Higgins as the title for his epic tome of 1833. He explained his work as *"An Attempt to Draw Aside the Veil of the Saitic Isis; or an Inquiry into the Origin of Languages, Nations and Religions."* His was indeed a revealing study of the ancient riddles, and I gladly count myself in his debt, but he fell short of drawing that veil aside. In fact, he failed even to understand its significance correctly.

Higgins was referring to an inscription on the ancient Temple of Isis at Säis that ended with the statement, "... No mortal man hath ever me unveiled." Certainly I do not claim to draw that particular veil aside; in fact, we shall see in due course that the mere passage of time has already accomplished what no mortal man ever could. But having discovered the elusive key to some of those puzzles I am able to draw aside the *figurative* veil that has obscured the beginnings of history these many years —thereby to shed light in areas where darkness has reigned heretofore.

Richard Petersen

Phoenix, Arizona
November, 1997

ACKNOWLEDGEMENTS

I AM PLEASED to express my gratitude to the librarians serving the Arizona collection at the Arizona State University for the generous help which they offered me on many occasions. I am also indebted to Minnabelle Laughlin of the Department of Anthropology for permitting me access to the Frank Midvale papers which are in her care. I am grateful as well to Tracy Meade, Curator of the Mesa Museum of History and Archaeology, for allowing me access to his "back room" where additional papers of Frank Midvale were still in storage, and also to Terry Hoagland, Curator of the Phoenix Museum of History, for his generous assistance and for allowing me access to his source materials.

I am grateful also to Mireya Ehlenberger for reading and reporting upon Father Kino's discussion of the comet of 1680. The Padre wrote Spanish with a thick German flavor so the task was more difficult than it promised to be at first glance.

The photographs included with the text are my own unless otherwise credited. One such exception required special effort on the part of Park Ranger Ramond Olivas, so I am particularly grateful to him, and to the National Park Service, for the reproduction of the de Vargas inscription at El Morro National Monument used here as Plate 6. I would also like to express my appreciation to the Arizona Historical Foundation for allowing me to reproduce the early photograph of the Casa Grande ruin in Plate 1, and to PARIS MATCH for permitting the use of their photograph of the restored moai in Plate 50.

In a slightly different vein I am pleased to thank Richard Potter for his generous gift of the sample shown here in Plates 57 and 58. It is a treasure indeed.

Next I must acknowledge my debt to Ted Holden, Peter Lamb, Walter Alter, Andrew MacRae and Wayne Throop for bringing to my attention the strange problem with *Pteranodon* and the sauropods, for which I propose an unexpected solution in Appendix E. Also I am indebted to Neil Steede for sharing with me some of his findings concerning the ruin at Tiahuanaco before the publication of his own report. I am also grateful to Joseph Mastropaolo for his continued support and helpful suggestions about the text. Likewise I am pleased to acknowledge the valuable editorial comments and advice of David Fields.

And finally, I am delighted above all else to remember the gracious help and encouragement of Roberta Gage, without which this work could not have been completed.

To Roberta

as a token

When folly is once taught it is very difficult to unteach it.

                                        Godfrey Higgins

Prologue:

A Brief Introduction

WILLIAM JAMES, the noted writer, once observed that "around and about the accredited and orderly facts of every science there ever floats a sort of dust-cloud of exceptional observances—an unresolved residuum of occurrences, minute and irregular, that always prove easier to ignore than to address forthrightly." His words are especially appropriate to some of the earth sciences where examples of such exceptional occurrences are commonplace. The great mammoths frozen in the tundra of Siberia offer a case in point; remarkably, some of them have residues of tropical forage still undigested in their stomachs. The ice ages constitute another such intractable riddle; presumably they came and went several times—*but only in the northern hemisphere,* not in the southern. Equally puzzling is evidence that the earth's magnetic field has reversed itself—and not only once but several times. This field is a deep mystery in its own right, but that it should have reversed direction at various times presents a riddle of higher order yet. In fact, wherever one probes beyond the dawn of history he uncovers an assortment of puzzles great and small that obscure our view of the ancient world like a thick cloud of dust indeed.

My goal in the following pages will be to dispel some of

that "dust" simply by addressing those unseemly riddles forthrightly for the first time ever. Heretofore, those who have vied in this arena first hobbled themselves with constraints that severely limited their vision and imagination. Thus hampered they stumbled over trifling impediments instead of stepping easily around them. Let us review briefly how this unlikely situation came into being.

I trace the beginning of the problem to that revolutionary movement in Eighteenth-century Europe that has come to be known as The Enlightenment. It took root amongst free-thinking intellectuals who chafed under dogmas imposed by the Church, and then it flowered with Newton's discovery of the law of gravitation. Whereas men had been told that God held the universe in his hands and caused it to operate, now they found that Nature obeyed certain fundamental laws inherently. By means of this understanding they were able to trace planets across the sky and predict the return of comets decades in advance. God had nothing to do with them. A few years later men found that strict laws also governed chemical reactions and the behavior of gases. Here again God played no part. Since Nature yielded to these laws then presumably she must yield to others as well; the rules she followed had only to be discovered, and then all the cosmos must give way to human understanding.

In due course a host of new sciences sprang up as first one aspect of nature and then another came under scrutiny with this vision in mind. Thus did the word "science" acquire a different meaning than before. Instead of signifying an ordered system of knowledge that could be distilled down into a few fundamental principles it came to include the *mere attempt* to understand nature in rational terms even when no distillation of principle was possible. And in fact the term is now applied to almost any excercise of the intellect which is played out in an arena where human reason reigns supreme. We call this arena 'The Laboratory'.

A Brief Introduction

As we know, the study of antiquity was among the first of the humanities to take on the form of science. In keeping with their rationalistic ideal Enlightened intellectuals abandoned the Biblical account of origins and set about to describe earth's development in terms of familiar processes operating in accord with the newly discovered laws of physics and chemistry. This aspiration, called the Principle of Uniformity, underlies all of modern geology. It assumes that the earth arrived at its present state "uniformly" over a prolonged period of time instead of quickly, as during episodes of great upheaval. More precisely, the Principle does not deny catastrophic events altogether, but it does insist that any such episodes must have conformed to the known laws of physics. Thus do geologists set aside any thought of the miraculous as an earth shaping mechanism, but of course in so doing they exclude from consideration any phenomenon whatever that fails to accord with human understanding. Evidently then, that "dust cloud of exceptional occurrences" having been excluded, the Laboratory does not encompass the full scope of Nature. It is instead only a *simplified model of nature* that may fall short in some circumstances.

That process of exclusion may take either of two forms. Some unseemly occurrences can be ignored outright, but those that cannot be ignored can be trivialized—that is, their *significance* can be ignored. As examples of this latter form of exclusion one might cite the dramatic atmospheric phenomena associated with the weather. In actual fact none of the atmospheric extremes, neither hurricanes, tornadoes, nor even the common thunderstorm can be understood in terms of normal physics. Let the reader note that these are not merely complex phenomena; they are deep mysteries in that their underlying mechanisms and the source of their energy are utterly unknown. Scientists suppose that their driving energy stems ultimately from the sun, but they cannot explain how it becomes concentrated and able to work such violence. Thus, if they were challenged to derive atmospheric behavior from the basic laws of physics scientists confined to The Laboratory would fail to envision any of the dramatic processes that we see routinely every day. They would even fail to anticipate the normal fall of rain.

But now let us recall that like-minded colleagues of those same scientists, also confined to The Laboratory, have devised a history of the earth that is similarly lacking in drama even though the very hills testify in anguish to earth-wrenching upheaval at some time—or times in the past. With the precedent of severe atmospheric phenomena in mind surely it is only a step to suppose that Nature may have still other weapons in her arsenal, likewise beyond the range of normal physics, by which she has formed the earth to its present state—not uniformly as geologists imagine, but intermittently during episodes of catastrophic upheaval. In the following pages, then, we undertake a search for such a weapon. And we shall find it, but to do so we shall be obliged to cast aside the *unnatural* constraints that researchers in this realm have hobbled themselves with in the past.

My interest in this subject began on reading Immanuel Velikovsky's notorious work, *Worlds in Collision*, published in 1950. As is well known, this book sparked a storm of controversy amongst academicians, some of whom took deliberate steps in an attempt to suppress it. His picture of events in the ancient world differed so remarkably from the current norm that it would have been ignored altogether if he had not presented compelling evidence to support it. Velikovsky's great offense was that he found a home for a number of facts that more conservative historians and scientists could not even bring themselves to acknowledge; mainstream academicians had consigned those embarrassing oddities to limbo, and they wanted very much for them to stay there.

According to his scenario a number of extraordinary natural disasters befell our planet in ancient times—disasters that toppled civilizations and left their marks on the very face

A Brief Introduction

of the earth. He identified the agent of this destruction as a comet. Thus far one could hardly quarrel with him; a collision between the earth and a comet would certainly wreak havoc on a large scale. But his picture required something other than an outright collision; it envisioned only a near miss, with the comet proceeding along its way, later to become upon aging the planet Venus. He then went on to postulate several catastrophic consequences of that near miss here on earth. Since my formal education had concentrated heavily on physics I had reason to be certain that he was largely wrong in his interpretion of the evidence, but his faulty conclusions in no way impeached the evidence itself. Assuredly some other interpretation of his data had to exist, and I found myself searching for it, more or less casually, taking my efforts only half seriously, during the years that followed.

Quite independently of this interest I had long been puzzled by a group of odd-looking hills within walking distance of my own home which by no stretch of the imagination could have been formed by processes acting conformably with the laws of physics. I was, therefore, confident to a certainty that the Uniformity Principle was not strictly valid—that otherwise unknown processes had at times been active in forming the earth. However, I could deduce nothing from the hills about the nature of those processes, and Velikovsky's evidence likewise remained very much of a puzzle to me. I did not think to connect the two riddles for many years.

But all of that changed one day as I was probing through the loess in Iowa. Thought to stem from the ice ages this silty deposit covers much of that state, as well as Nebraska and Kansas. I had previously pored over the geological literature on the subject so I was not prepared for surprises, but surprised I was indeed. For there before my eyes and in my very hands were the most extraordinary objects that one could possibly imagine. They were so unexpected that their meaning escaped me at first, but as surprise gave way to reflection they told a remarkable story very clearly. That story, and an interpretation of those strange hills near my home, form the core of this work. At first sight they appear to be two entirely different stories, but they converge to a common theme in the end.

That theme is stark catastrophe—catastrophe, moreover, that does not conform to the Uniformity Principle. We shall therefore be obliged to step outside of The Laboratory in order to understand it even superficially, but having done so we shall be able to resolve a great many riddles from antiquity that are otherwise wholly intractable; as one case in point we shall discover where Velikovsky went astray in his deductions about the planet Venus. Likewise the awful significance of that unresolved residuum of irregular occurrences that William James spoke about will become apparent. Although these new insights may engender some surprise at first I anticipate that most readers will be delighted with them and will breathe a sigh of relief that the wonderful fullness of nature can now be acknowledged forthrightly at last. Concerning the social and philosophical implications of these findings I shall offer a few words of personal comment in conclusion.

How best to enter for the first time that realm beyond The Laboratory door is not easy to decide. Certainly the route laid out in the following pages is not the only one possible, but it is at least easy to travel and proves fruitful in the end though progress may seem slow at first. In fact, it is essentially the same path that I myself followed as I entered this daunting new territory for the first time.

Now then, in keeping with the principle that the best place to start is at the beginning, we address our attention first to a train of events that took place shortly after the Spanish conquest of Mexico; they define the setting for the more recent of those two principal stories. At issue here is the identity of a sprawling Indian metropolis situated many miles to the north of Mexico City.

Chapter 1:

THE GREAT CITIES

WHAT FOLLOWS HAS all the ingredients of a classic mystery story. Laws have been broken; witnesses have lied; puzzling clues abound; evidence has been tampered with; great cities have vanished, and the authorities are baffled.

Challenged with this real-life mystery we might cast ourselves in the role of a private detective who has been consulted on the case. To set the scene, let us suppose that our prospective client is a young woman, perhaps thirty years of age, who has just been ushered into our lodgings from a thick wintry fog outside. She apologizes for calling without an appointment and then explains, "I resolved to consult with you, Mr. Holmes, because I've been haunted of late by a gnawing suspicion—a suspicion that I am the victim of an elaborate hoax."

Our interest quickens, and after being seated by the fire our visitor continues, "It causes me no end of embarrassment in my line of studies, Mr. Holmes, although it was not aimed at me personally. Indeed, the curtain has long since fallen on the stage where the events in question played out. Even the play itself has been largely forgotten, but I would ask you to apply your great powers of deduction to the shadows that remain from those olden times and explain what actually happened. I've pondered the problem at length myself, of course, but can claim no progress against it whatever."

Hardly does our caller utter these words when she pales suddenly and leans forward upon her knees for support. Ever solicitous of the gentler sex Dr. Watson offers her a snifter of brandy which she accepts gladly, explaining that the crossing from America was excessively arduous and she has not yet recovered from the ordeal. Taking advantage of the respite we fill our pipe from the slipper nearby, and when the color has returned to her cheeks we invite the young lady to proceed. She pauses for a moment to collect her thoughts and then continues.

"The events that you must know about, Mr. Holmes, began to unfold in Mexico shortly after the Spanish Conquest. That took place in the year 1521, as you may know. Well, the Spaniards had hardly gotten settled in their new colony when rumors began to circulate about other great cities far to the north. In fact, Nuño de Guzmán, President of the first *Audiencia* of New Spain (Mexico), preserved the account of an Indian servant who said that as a boy he had once traveled to such cities with his father, who was a trader. According to him they were situated forty days' travel north of his home village and compared in size with Mexico City itself.

One might suppose that those enthusiastic rumors originated with primitive nomads who were entirely unfamiliar with permanent housing; such people might describe even modest pueblos in glowing terms, but we see that this was not the case at all. The witness here was at ease in Mexico City so he was well acquainted with urban culture, and the President himself took the man at his word. Indeed, de Guzmán assembled an army and in 1530 set out to find and conquer those cities that the Indian had described. He could not cope with the mountainous terrain so he did not get very far, but he did carve out a substantial province for himself which he called New Galicia and of which he became the first Governor. However de Guzmán's barbarous treatment of the natives, both during the

The Great Cities

expedition and afterwards, eventually got him into serious trouble as we shall presently see.

Our next word of the wonderful cities comes from four gaunt men discovered by a military patrol on the outskirts of de Guzmán's new territory in the year 1536. They were Álvar Nuñez Cabeza de Vaca*, Andrés Dorantes, Alonzo del Castillo Maldonado and an Arabian Negro named Estéban. The latter was a slave, the liege of Dorantes. Those four were the sole survivors of the ill-fated Narváez expedition that had set out to conquer Florida eight years before.

That campaign proved a misadventure from the beginning. The army was not prepared for guerrilla warfare in the wilds so they became disorganized and did poorly against the Indians. To make matters worse their ships were lost in storms so they could not escape. Disease also took its toll. Many who did not succumb outright became too weak to resist and were executed by their foes as they were found, but for some reason those four were taken as slaves instead. After suffering unspeakable privation for six years, however, they managed to elude their masters and to work their way westward and south into Mexico where they were finally rescued. Near the end of their journeys the Indians along the way had spoken glowingly of marvelous cities to the north, and they even displayed various tangible items which they said had come from there.

In due course Alvar Nuñez and the others found themselves in Mexico City where they were received with great celebrations. Bull fights were held in their honor, and they were welcomed by Hernando Cortéz himself. In their report to the Viceroy, Don Antonio de Mendoza, they naturally mentioned those large and powerful cities of which they had been told repeatedly along the way.

---

* *The man's name was Álvar Nuñez. His cognomen, Cabeza de Vaca, (Head of a Cow) was bestowed upon an ancestor by the King in honor of an act of heroism that involved a cow's head.*

Mendoza was greatly interested in those reports, and he was anxious to determine if they were true. With this object in mind he tried to persuade the four, any or all, to retrace their route and attempt to find those cities, but most had other plans. Dorantes and Castillo wished to marry and settle down to a more tranquil life, while Álvar Nuñez wanted to return to Spain. The Negro was eager to go, but because of his station he could not be given the responsibility for such an important undertaking; someone else would have to be found to lead the expedition. In order to be helpful Dorantes ceded Estéban to the Viceroy that he might be available for such an effort if it should ever materialize. Three years passed before Mendoza did finally assemble a scouting party to investigate, and he selected a priest to lead it. There has been much learned conjecture about his reasons for making such a choice. They may have been largely political, as some suggest, or he may simply have known the man and had confidence in him. In any case, he chose to lead the mission a Franciscan priest by the name of Marcos de Niza. Little is known of Fray Marcos save that he lived for a time in Nice (de Niza) and was with Pizarro during the conquest of Peru. His father provincial attested that he was highly esteemed for both his character and learning—not only in theology but also in cosmography and navigation.

Now let us note well that this was an official undertaking; the Viceroy himself had ordered the expedition, and it was a matter of the utmost gravity in the eyes of all concerned. He presented Fray Marcos with detailed written instructions, and Marcos in turn was obliged to file a written receipt stating that he understood his orders and promised faithfully to obey them. In effect, Marcos was to be the eyes of the King in that new territory. It was a perfectly legitimate assignment, so we have every obvious reason to believe that the priest undertook his mission in all good faith and that he filed a true and accurate report upon its conclusion. All of those documents are matters

The Great Cities

of public record to this day.

We can do no better than read the account of that trek as Fray Marcos wrote it. Some details may seem remote from our central problem—that of locating the cities, but they all shed light on the question of his truthfulness. In effect, we shall be able to take a measure of the man from his writing. And furthermore, by reading the full account we shall learn not only what he said but also what he did not say. The following

FIGURE 1

translation of his report into English was prepared by Dr. Percy Baldwin for the New Mexico Historical Society [6], and it is reproduced here in its entirety with only two minor variations. Baldwin anglicized the Negro's name (Stephen of Dorantes) while retaining the name of the cities in its original form—with an accent on the first syllable. There can be no doubt that Marcos wrote the name as it was told to him so Cíbola is surely correct. But here the personal name is retained as in the Spanish, while the place name is anglicized; that is, the accent is omitted. Also, the paragraphs have been numbered so they can be referred to easily.

Fray Marcos refers many times to crossing "deserts" during his trek, but one should be aware that the Spanish word, *desplobado,* is used to denote any unpopulated region whatever so it is not strictly the equal of the English desert. One should also know that the Spanish league was taken to be the distance a horse would normally walk in an hour, and it is usually reckoned as 3.1 miles. Five such leagues, or about 15.5 miles, were considered the normal day's travel on foot.

Figure 1 is a sketch map of the territory that we shall be considering here and in the chapters to follow. Apart from Mexico City itself, only two of the centers shown were known to be in existence in 1539, namely Compostella and Culiacán, which was called San Miguel at the time.

The reader cannot fail to notice that New Galicia had a new Governor by the time Fray Marcos set out on his journey. De Guzmán's continued abuse of the Indians had caused him to be stripped of office and thrown into prison. The new Governor, Francisco Vázquez de Coronado, was a young man, well-born and well-married, of whom we shall hear more later. Now let us turn our attention to what purports to be eyewitness testimony for the living fact of those great cities—an account that was written more than eighty years before the Pilgrim Fathers anchored the "Mayflower" and set their feet upon the New World!

THE REPORT

OF Fray Marcos de Niza

(1) With the aid and favor of the most holy Virgin Mary, our Lady, and of our seraphic father, St. Francis, I, Fray Marcos de Niza, a professed religious of the order of St. Francis, in fulfillment of the instructions above given of the most illustrious lord, Don Antonio de Mendoza, viceroy and governor for H. M. of New Spain, left the town of San Miguel, in the province of Culiacán, on Friday, March 7th, 1539. I took with me as companion Friar Honoratus and also Estéban de Dorantes, a negro, and certain Indians, which the said Lord Viceroy bought for the purpose and set at liberty. They were delivered to me by Francisco de Coronado, governor of New Galicia, along with many other Indians from Petatlan and from the village of Cuchillo, situated about fifty leagues from the said town. All these came to the valley of Culiacán, manifesting great joy, because it had been certified to them that the Indians were free, the said governor having sent in advance to acquaint them of their freedom and to tell them that it was the desire and command of H. M. that they should not be enslaved nor made war upon nor badly treated.

(2) With this company as stated, I took my way toward the town of Petatlan, receiving much hospitality and presents of food, roses and other such things; besides which, at all the stopping places where there were no people, huts were constructed for me of mats and branches. In this town of Petatlan I stayed three days, because my companion, Friar Honoratus, fell sick. I found it advisable to leave him there and, conformably with the instructions given to me, I followed the way in which I was guided, though unworthy, by the Holy Ghost. There went with me Estéban de Dorantes, the negro, some of the freed Indians and

many people of that country. I was received everywhere I went with much hospitality and rejoicing and with triumphal arches. The inhabitants also gave me what food they had, which was little, because they said it had not rained for three years, and because the Indians of that territory think more of hiding than of growing crops, for fear of the Christians of the town of San Miguel, who up to that time were accustomed to make war upon and enslave them. On all this road, which would be about 25 or 30 leagues beyond Petatlan, I did not see anything worthy of being set down here, except that there came to me some Indians from the island visited by the Marquess of Valle, and who informed me that it was really an island and not, as some think, part of the mainland. I saw that they passed to and from the mainland on rafts and that the distance between the island and the mainland might be half a sea league, rather more or less. Likewise there came to see me Indians from another larger and more distant island by whom I was told that there were thirty other small islands, inhabited, but with poor food excepting two, which they said had maize. These Indians wore suspended from their necks many shells of the kind which contain pearls; I showed them a pearl which I carried for sample and they told me that there were some in the islands, but I did not see any.

(3)   I took my way over a desert for four days and there went with me some Indians from the islands mentioned as well as from the villages which I left behind, and at the end of the desert I found some other Indians, who were astonished to see me, as they had no news of Christians, having no traffic with the people on the other side of the desert. These Indians made me very welcome, giving me plenty of food, and they endeavored to touch my clothes, calling me Sayota, which means in their language, "man from heaven." I made them understand, the best I could by my interpreters, the content of my instructions, namely, the knowledge of our Lord in heaven and of H. M. on earth. And always, by all the means that I could, I sought to learn about a country with numerous towns and a people of a higher culture than those I was encountering, but I had no news except that they told me that in the country beyond, four or five days' journey thence, where the chains of mountains ended, there was an extensive and level open tract,

The Great Cities

in which they told me there were many and very large towns inhabited by a people clothed with cotton. When I showed them some metals which I was carrying, in order to take account of the metals of the country, they took a piece of gold and told me that there were vessels of it among the people of the region and that they wear certain articles of that metal suspended from their noses and ears, and that they had some little blades of it, with which they scrape and relieve themselves of sweat. But as this tract lies inland and my intention was to stay near the coast, I determined to leave it till my return, because then I would be able to see it better. And so I marched three days through a country inhabited by the same people, by whom I was received in the same manner as by those I had already passed. I came to a medium-sized town named Vacapa, where they made me a great welcome and gave me much food, of which they had plenty, as the whole land is irrigated. From this town to the sea is forty leagues. As I found myself so far away from the sea, and as it was two days before Passion Sunday, I determined to stay there until Easter, to inform myself concerning the islands of which I said above that I had news. So I sent Indian messengers to the sea, by three ways, whom I charged to bring back to me people from the coast and from some of the islands, that I might inform myself concerning them. In another direction I sent Estéban de Dorantes, the negro, whom I instructed to take the route toward the north for fifty of sixty leagues to see if by that way he might obtain an account of any important thing such as we were seeking. I agreed with him that if he had any news of a populous, rich and important country he should not continue further but should return in person or send me Indians with a certain signal which we arranged, namely, that if it were something of medium importance he should send me a white cross of a hand's breadth, if it were something of great importance, he should send me one of two hands' breadth, while if were bigger and better than New Spain, he should send me a great cross. And so the said negro Estéban departed from me on Passion Sunday after dinner, whilst I stayed in the town, which as I say is called Vacapa. (4) In four days' time there came messengers from Estéban with a very great cross, as high as a man, and they told me on Estéban's behalf that I should immediately come and follow him, because he had met people who gave him an account of the greatest country in the world, and that he had Indians who had been there, of whom he sent me one. This man told me so many wonderful things about the country, that I forbore to believe them until I should have seen them, or should have more certitude of the matter. He told me that it was thirty days' journey from where Estéban was staying to the first city of the country, which was named Cibola. As it appears to be worth while to put in this paper what this Indian, whom Estéban sent me, said concerning the country, I will do so. He asserted that in the first province there were seven very great cities, all under one lord, that the houses, constructed of stone and lime, were large, that the smallest were of one story with a terrace above, that there were others of two and three stories, whilst that of the lord had four, and all were joined under his rule. He said that the doorways of the principal houses were much ornamented with turquoises, of which there was a great abundance, and that the people of those cities went very well clothed. He told me many particulars, not only of the seven cities but of other provinces beyond them, each one of which he said was much bigger than that of the seven cities. That I might understand the matter as he knew it, we had many questions and answers and I found him very intelligent.

(5)   I gave thanks to Our Lord, but deferred my departure after Estéban de Dorantes, thinking that he would wait for me, as I had agreed with him, and also because I had promised the messengers whom I had sent to the sea that I would wait for them, for I proposed always to treat with good faith the people with whom I came in contact. The messengers returned on Easter Sunday, and with them people from the coast and from two islands, which I knew to be the islands above mentioned, and which, as I already knew, are poor of food, though populated. These people wore shells on their foreheads and said that they contain pearls. They told me that there were thirty-four islands near to one another, whose names I am setting down on another paper, where I give the names of the islands and towns. The people of the coast say that they, as well as the people of the islands, have little food, and that they traffic with one another by means of rafts. The coast trends

The Great Cities

almost directly toward the north. These Indians of the coast brought to me shields of oxhide, very well fashioned, big enough to cover them from head to foot, with some holes above the handle so that one could see from behind them; they are so hard, that I think that a bullet would not pass through them. The same day there came to me three of those Indians known as Pintados, with their faces, chests and arms all decorated; they live over toward the east and their territory borders on those near the seven cities. They told me that, having had news of me, they had come to see me and among other things they gave me much information concerning the seven cities and provinces that the Indian sent by Estéban had told me of, and almost in the same manner as he. I therefore sent back the coast people, but two Indians of the islands said they would like to go with me seven or eight days.

(6)   So with them and the three Pintados already mentioned, I left Vacapa on the second day of the Easter festival, taking the same road that Estéban had followed. I had received from him more messengers, with another big cross as big as the first which he sent, urging me to hurry and stating that the country in question was the best and greatest of which he had ever heard. These messengers gave me, individually, the same story as the first, except that they told me much more and gave me a clearer account. So for that day, the second of Easter, and for two more days I followed the same stages of the route as Estéban had; at the end of which I met the people who had given him news of the seven cities and of the country beyond. They told me that from there it was thirty days' journey to the city of Cibola, which is the first of the seven. I had an account not from one only, but from many, and they told me in great detail the size of the houses and the manner of them, just as the first ones had. They told me that beyond these seven cities there were other kingdoms named Marata, Acus and Totonteac. I desired very much to know for what they went so far from their homes and they told me that they went for turquoises, cowhides and other things, that there was a quantity of these things in that town. Likewise I asked what they exchanged for such articles and they told me the sweat of their brows and the service of their persons, that they went to the first city, which is called Cibola, where they served in

digging the ground and performing other work, for which work they are given oxhides of the kind produced in that country, and turquoises. The people of this town all wear good and beautiful turquoises hanging from their ears and noses and they say that these jewels are worked into the principal doors of Cibola. They told me that the fashion of clothing worn in Cibola is a cotton shirt reaching to the instep, with a button at the throat and a long cord hanging down, the sleeves of the shirts being the same width throughout their length; it seems to me this would resemble the Bohemian style. They say that those people go girt with belts of turquoises and that over these shirts some wear excellent cloaks and others very well dressed cowhides, which are considered the best clothing, and of which they say there is a great quantity in that country. The women likewise go clothed and covered to the feet in the same manner.

(7) These Indians received me very well and took great care to learn the day of my departure from Vacapa, so that they might furnish me on the way with victuals and lodgings. They brought me sick persons that I might cure them and they tried to touch my clothes; I recited the Gospel over them. They gave me some cowhides so well tanned and dressed that they seemed to have been prepared by some highly civilized people, and they all said that they came from Cibola.

(8) The next day I continued my journey, taking with me the Pintados, who wished not to leave me. I arrived at another settlement, where I was very well received by its people, who also endeavored to touch my clothing. They gave me information concerning the country whither I was bound as much in detail as those I had met before, and they told me that some persons had gone from there with Estéban de Dorantes, four or five days previously. Here I found a great cross which Estéban had left for me, as a sign that the news of the good country continually increased, and he had left word for me to hurry and that he would wait for me at the end of the first desert. Here I set up two crosses and took possession, according to my instructions, because that country appeared to me better than that which I had already passed and hence it was fitting to perform the acts of possession.

(9) In this manner I traveled five days, always finding people,

The Great Cities

who gave me a very hospitable reception, many turquoises and cowhides and the same account of the country. They all spoke to me right away of Cibola and that province as people who knew that I was going in search of it. They told me how Estéban was going forward, and I received from him messengers who were inhabitants of that town and who had been some distance with him. He spoke more and more enthusiastically of the greatness of the country and he urged me to hurry. Here I learned that two days' journey thence I would encounter a desert of four days' journey, in which there was no provision except what was supplied by making shelters for me and carrying food. I hurried forward, expecting to meet Estéban at the end of it, because he had sent me word that he would await me there.

(10)   Before arriving at the desert, I came to a green, well watered settlement, where there came to meet me a crowd of people, men and women, clothed in cotton and some covered with cowhides, which in general they consider a better dress material than cotton. All the people of this town wear turquoises hanging from their noses and ears; these ornaments are called cacona. Among them came the chief of the town and his two brothers, very well dressed in cotton, encaconados, and each with a necklace of turquoises around his neck. They brought to me a quantity of game—venison, rabbits and quail—also maize and meal, all in great abundance. They offered me many turquoises, cowhides, very pretty cups and other things, of which I accepted none, for such was my custom since entering the country where we were not known. And here I had the same account as before of the seven cities and the kingdoms and provinces as I have related above. I was wearing a garment of dark woolen cloth, of the kind called Saragossa, which was given to me by Francisco Vázquez de Coronado, governor of New Galicia. The chief of the village and other Indians touched it with their hands and told me that there was plenty of that fabric in Totonteac, and that the natives of that place were clothed with it. At this I laughed and said it could not be so, that it must be garments of cotton which those people wore. Then they said to me: "Do you think that we do not know that what you wear and what we wear is different? Know that in Cibola the houses are full of that material which we are wearing, but

in Totonteac there are some small animals from which they obtain that with which they make a fabric like yours." This astonished me, as I had not heard of any such thing previously, and I desired to inform myself more particularly about it. They told me that the animals are of the size of the Castilian greyhounds which Estéban had with him; they said there were many of them in Totonteac. I could not guess what species of animals they might be.

(11)   The next day I entered into the desert and at the place where I had to go for dinner I found huts and food enough by the side of a watercourse. At night I found cabins and food again and so it was for the four days that I traveled through this desert. At the end of them, I entered a very well populated valley and at the first town many men and women came with food to meet me. They all wore many turquoises suspended from their noses and ears, and some wore necklaces of turquoise, like those which I said were worn by the chief of the town on the other side of the desert, and his brothers, except that they only wore one string, while these Indians wore three or four. They were dressed in very good cloaks of ox leather. The women likewise wore turquoises in their noses and ears and very good petticoats and blouses. Here they had as much information of Cibola, as in New Spain they have of Mexico and in Peru of Cuzco. They described in detail the houses, streets and squares of the town, like people who had been there many times, and they were wearing various objects brought from there, which they had obtained by their services, like the Indians I had previously met. I said to them that it was not possible that the houses should be in the manner which they described to me, so to make me understand they took earth and ashes and mixed them with water, and showed how the stone is placed and the edifice reared, placing stone and mortar till the required height is reached. I asked them if the men of that country had wings to climb those stories; they laughed and explained to me a ladder as well as I could do, and they took a stick and placed it over their heads and said it was that height from story to story. Here I was also given an account of the woolen cloth of Totonteac, where they say the houses are like those at Cibola, but better and bigger, and that it is a very great place and has no limit.

The Great Cities

(12)   Here I learned that the coast turns to the west, almost at a right angle, because until I reached the entrance of the first desert which I passed, the coast always trended toward the north. As it was very important to know the direction of the coast, I wished to assure myself and so went to look out and I saw clearly that in latitude 35 degrees it turns to the west. I was not less pleased at this discovery than at the good news I had of the country.

(13)   So I turned to follow my route and was in that valley five days. It is so thickly populated with fine people and so provided with food that there would be enough to supply more than three hundred horse. It is all watered and is like a garden. There are villages at every half or quarter league or so. In each of them I had a very long account of Cibola, and they spoke to me in detail about it, as people who went there each year to earn their living. Here I found a man who was a native of Cibola. He told me he had fled from the governor whom the lord had placed there in Cibola—for the lord of these seven cities lives and has his residence in one of them, which is called Ahacus, and in the others he has placed persons who command for him. This citizen of Cibola is a man of good disposition, somewhat old and much more intelligent than the natives of the valley and those I had formerly met; he told me that he wished to go with me so that I might procure his pardon. I interrogated him carefully and he told me that Cibola is a big city, that it has a large population and many streets and squares, and that in some parts of the city there are very great houses, ten stories high in which the chiefs meet on certain days of the year. He corroborated what I had already been told, that the houses are constructed of stone and lime, and he said that the doors and fronts of the principal houses are of turquoise; he added that the others of the seven cities are similar, though some are bigger, and that the most important is Ahacus. He told me that towards the southeast lay a kingdom called Marata, in which there used to be many very large towns, having the same kind of houses built of stone and with several stories; that this kingdom had been, and still was, at war with the lord of the seven cities; that by this war Marata had been greatly reduced in power, although it was still independent and continued the war.

(14)   He likewise told me that to the southeast there is a kingdom named Totonteac, which he said was the biggest, most populous, and the richest in the world, and that there they wore clothes made of the same stuff as mine, and others of a more delicate material obtained from the animals of which I had already had a description; the people were highly cultured and different from those I had hitherto seen. He further informed me that there is another province and very great kingdom, which is called Acus—for there are Ahacus and Acus; Ahacus, with the aspiration, is one of the seven cities, and the most important one, and Acus, without the aspiration, is a kingom and province by itself.

(15)   He corroborated what I had been told concerning the clothes worn in Cibola and added that all the people of that city sleep in beds raised above the floor, with fabrics and with tilts above to cover the beds. He said that he would go with me to Cibola and beyond, if I desired to take him along. I was given the same account in this town by many other persons, though not in such great detail.

(16)   I traveled in this valley three days and the natives made for me all the feasts and rejoicings that they could. Here in this valley I saw more than two thousand oxhides, extremely well cured; I saw a very large quantity of turquoises and necklaces thereof, as in the places I had left behind, and all said that they came from the city of Cibola. They know this place as well as I would know what I hold in my hands, and they are similarly acquainted with the kingdoms of Marata, Acus and Totonteac. Here in this valley they brought to me a skin, half as big again as that of a large cow, and told me that it was from an animal which has only one horn on its forehead and that this horn is curved towards its chest and then there sticks out a straight point, in which they said there was so much strength, that no object, no matter how hard, could fail to break when struck with it. They averred that there were many of these animals in that country. The color of the skin is like that of the goat and the hair is as long as one's finger.

(17)   Here I had messengers from Estéban, who told me on his behalf that he was then entering the last desert, and the more cheerfully, as he was going more assured of the country; and he sent to me to say

The Great Cities

that, since departing from me, he had never found the Indians out in any lie, but up to that point had found everything as they had told him and so he thought to find that beyond. And so I held it for certain, because it is true, that from the first day I had news of the city of Cibola, the Indians had told me of everything that till then I had seen, telling me always what town I would find along the road and the numbers of them and, in the parts where there was no population, showing me where I would eat and sleep, without erring in one point. I had then marched, from the first place where I had news of the country, one hundred and twelve leagues, so it appears to me not unworthy to note the great truthfulness of these people. Here in this valley, as in the other towns before, I erected crosses and performed the appropriate acts and ceremonies, according to my intructions. The natives of this town asked me to stay with them three or four days, because there was a desert four leagues thence, and from the beginning of it to the city of Cibola would be a march of fifteen days and they wished to put up food for me and to make the necessary arrangements for it. They told me that with the negro Estéban there had gone more than three hundred men to accompany him and carry food, and that many wished to go with me also, to serve me and because they expected to return rich. I acknowledged their kindness and asked that they should get ready speedily, because each day seemed to me a year, so much I desired to see Cibola. And so I remained three days without going forward, during which I continually informed myself concerning Cibola and all the other places. In doing so I took the Indians aside and questioned each one by himself, and all agreed in their account and told me the number of the people, the order of the streets, the size of the houses and the fashion of the doorways, just as I had been told by those before.

(18) After the three days were past, many people assembled to go with me, of whom I chose thirty chiefs, who were very well supplied with necklaces of turquoises, some of them wearing as many as five or six strings. With these I took the retinue necessary to carry food for them and me and started on my way. I entered the desert on the ninth day of May. On the first day, by a very wide and well traveled road, we arrived for dinner at a place where there was water, which the

Indians showed to me, and in the evening we came again to water, and there I found a shelter which the Indians had just constructed for me and another which had been made for Estéban to sleep in when he passed. There were some old huts and many signs of fire, made by people passing to Cibola over this road. In this fashion I journeyed twelve days, always very well supplied with victuals of venison, hares, and partridges of the same color and flavor as those of Spain, although rather smaller.

(19) At this juncture I met an Indian, the son of one of the chiefs who were journeying with me, who had gone in the company with the negro Estéban. This man showed fatigue in his countenance, had his body covered with sweat, and manifested the deepest sadness in his whole person. He told me that, at a day's march before coming to Cibola, Estéban, according to his custom, sent ahead messengers with his calabash, that they might know he was coming. The calabash was adorned with some rows of rattles and two feathers, one white and one red. When they arrived at Cibola, before the person of the lord's representative in that place, and gave him the calabash, as soon as he took it in his hands and saw the rattles, with great anger he flung it on the ground and told the messengers to be gone forthwith, that he knew what sort of people these were, and that the messengers should tell them not to enter the city, as if they did so he would put them to death. The messengers went back and told Estéban what had passed. He said to them that that was nothing, that those who showed themselves irritated received him the better. So he continued his journey till he arrived at the city of Cibola, where he found people who would not consent to let him enter, who put him in a big house which was outside the city, and who at once took away from him all that he carried, his articles of barter and the turquoises and other things which he had received on the road from the Indians. They left him that night without giving anything to eat or drink either to him or to those that were with him. The following morning my informant was thirsty and went out of the house to drink from a nearby stream. When he had been there a few moments he saw Estéban fleeing away, pursued by the people of the city and they killed some of those who were with him. When this

The Great Cities

Indian saw this he concealed himself and made his way up the stream, then crossed over and regained the road of the desert.

(20)   At these tidings, some of the Indians who were with me commenced to weep. As for myself, the wretched news made me fear I should be lost. I feared not so much to lose my life as not to be able to return to give a report of the greatness of the country, where God, Our Lord, might be so well served and his holy faith exalted and the royal domains of H. M. extended. In these circumstances I consoled them as best I could and told them that one ought not to give entire credence to that Indian, but they said to me with many tears that the Indian only related what he had seen. So I drew apart from the Indians to commend myself to Our Lord and to pray Him to guide this matter as He might best be served and to enlighten my mind. This done, I returned to the Indians and, with a knife, cut the cords of the packages of dry goods and articles of barter which I was carrying with me and which till then I had not touched nor given away any of the contents. I divided up the goods among all those chiefs and told them not to fear and to go along with me, which they did.

(21)   Continuing our journey, at a day's march from Cibola, we met two other Indians, of those who had gone with Estéban, who appeared bloody and with many wounds. At this meeting, they and those that were with me set up such a crying, that out of pity and fear they also made me cry. So great was the noise that I could not ask about Estéban nor of what had happened to them, so I begged them to be quiet that we might learn what had passed. They said to me: "How can we be quiet, when we know that our fathers, sons, and brothers who were with Estéban, to the number of more than three hundred men, are dead? And we no more dare go to Cibola, as we have been accustomed." Nevertheless, as well as I could, I endeavored to pacify them and to put off their fear, although I myself was not without need of someone to calm me. I asked the wounded Indians concerning Estéban and as to what had happened. They remained a short time without speaking a word, weeping along with those of their towns. At last they told me that when Esteban arrived at a day's journey from Cibola, he sent his messengers with his calabash to the lord of Cibola to announce his

arrival and that he was coming peacefully and to cure them. When the messengers gave him the calabash and he saw the rattles, he flung it furiously on the floor and said: "I know these people; these rattles are not of our style of workmanship; tell them to go back immediately or not a man of them will remain alive." Thus he remained very angry. The messengers went back sad, and hardly dared to tell Estéban of the reception they had met. Nevertheless they told him and he said that they should not fear, that he desired to go on, because, although they answered him badly, they would receive him well. So he went and arrived at the city of Cibola just before sunset, with all his company, which would be more than three hundred men, besides many women. The inhabitants would not permit them to enter the city, but put them in a large and commodious house outside the city. They at once took away from Estéban all that he carried, telling him that the lord so ordered. "All that night," said the Indians, "they gave us nothing to eat nor drink. The next day, when the sun was a lance-length high, Estéban went out of the house and some of the chiefs with him. Straightaway many people came out of the city and, as soon as he saw them, he began to flee and we with him. Then they gave us these arrow-strokes and cuts and we fell and some dead men fell on top of us. Thus we lay till nightfall, without daring to stir. We heard loud voices in the city and we saw many men and women watching on the terraces. We saw no more of Estéban and we concluded that they had shot him with arrows as they had the rest that were with him, of whom there escaped only us."

(22)   In view of what the Indians had related and the bad outlook for continuing my journey as I desired, I could not help but feel their loss and mine. God is witness of how much I desired to have someone of whom I could take counsel, for I confess I was at a loss what to do. I told them that Our Lord would chastize Cibola and that when the Emperor knew what had happened he would send many Christians to punish its people. They did not believe me, because they say that no one can withstand the power of Cibola. I begged them to be comforted and not to weep and consoled them with the best words I could muster, which would be too long to set down here. With this

The Great Cities

I left them and withdrew a stone's throw or two apart, to commend myself to God, and remained thus an hour and a half. When I went back to them, I found one of my Indians, named Mark, who had come from Mexico, and he said to me: "Father, these men have plotted to kill you, because they say that on account of you and Estéban their kinsfolk have been murdered, and that there will not remain a man or woman among them all who will not be killed." I then divided among them all that remained of dry stuffs and other articles, in order to pacify them. I told them to observe that if they killed me they would do me no harm, because I would die a Christian and would go to heaven, and that those who killed me would suffer for it, because the Christians would come in search of me, and against my will would kill them all. With these and many other words I pacified them somewhat, although there was still high feeling on account of the people killed. I asked that some of them should go to Cibola, to see if any other Indians had escaped and to obtain some news of Esteban, but I could not persuade them to do so. Seeing this, I told them that, in any case, I must see the city of Cibola, and they said that no one would go with me. Finally, seeing me determined, two chiefs said that they would go with me.

(23)   With these and with my own Indians and interpreters, I continued my journey till I came within sight of Cibola. It is situated on a level stretch on the brow of a roundish hill. It appears to be a very beautiful city, the best that I have seen in these parts; the houses are of the type that the Indians described to me, all of stone, with their stories and terraces, as it appeared to me from a hill whence I could see it. The town is bigger than the city of Mexico. At times I was tempted to go to it, because I knew that I risked nothing but my life, which I offered to God the day I commenced the journey; finally I feared to do so, considering my danger and that if I died, I would not be able to give an account of this country, which seems to me to be the greatest and best of the discoveries. When I said to the chiefs who were with me, how beautiful Cibola appeared to me, they told me that it was the least of the seven cities, and that Totonteac is much bigger and better than all the seven, and that it has so many houses and people that there is no end to it. Viewing the situation of the city, it occurred to me to call

that country the new kingdom of St. Francis, and there, with the aid of the Indians, I made a big heap of stones and on top of it I placed a small slender cross, not having the materials to construct a bigger one. I declared that I placed that cross and landmark in the name of Don Antonio de Mendoza, viceroy and governor of New Spain and the Emperor, our lord, in the sign of possession, in conformity with my instructions. I declared that I took possession there of all the seven cities and of the kingdoms of Totonteac and Acus and Marata, and that I did not go to them in order that I might return to give an account of what I had done and seen.

(24)   Then I started back, with much more fear than food, and went to meet the people whom I had left behind, with the greatest haste I could make. I overtook them after two days' march and went with them till we had passed the desert and arrived at their home. Here I was not made welcome, as previously, because the men, as well as the women, indulged in much weeping for the persons killed at Cibola. Without tarrying, I hastened in fear from that people and that valley. The first day I went ten leagues, then I went eight and again ten leagues, without stopping till I had passed the second desert.

(25)   On my return, although I was not without fear, I determined to approach the open tract, situated at the end of the mountain ranges, of which I said above that I had some account. As I came near, I was informed that it is peopled for many days' journey towards the east, but I dared not enter it, because it seemed to me that we must go to colonize and to rule that other country of the seven cities and the kingdoms I have spoken of, and that then one could see it better. So I forbore to risk my person and left it alone to give an account of what I had seen. However, I saw from the mouth of the [valley] seven moderate-sized towns at some distance, and further a very fresh valley of very good land, whence rose much smoke. I was informed that there is much gold in it and that the natives of it deal in vessels and jewels for the ears and little plates with which they scrape themselves to relieve themselves of sweat, and that these people will not consent to trade with those of the other part of the valley; but I was not able to learn the cause for this. Here I placed two crosses and took possession of all this plain

The Great Cities

and valley in the same manner as I had done with the other possessions, according to my instructions. From there I continued my return journey, with all the haste I could, till I arrived at the town of San Miguel, in the province of Culiacán, expecting to find there Francisco Vázquez de Coronado, governor of New Galicia. As I did not find him there, I continued my journey to the city of Compostella, where I found him. From there I immediately wrote word of my coming to the most illustrious lord, the viceroy of New Spain, and to our father provincial, Friar Antonio, of Ciudad-Rodrigo, asking him to send me orders what to do.

(26)   I omit here many particulars which are not pertinent; I simply tell what I saw and what was told me concerning the countries where I went and those of which I was given information, in order to make a report to our father provincial, that he may show it to the father of our order, who may advise him, or to the council of the order, at whose command I went, that they may give it to the most illustrious lord, the viceroy of New Spain, at whose request they sent me on the journey.

Now as straightforward as this document seems to be, it has been a bone of contention for centuries. Does it have the "ring of truth" about it as some have said? Or are the important points mere fabrications—is it in fact of tissue of lies as most have judged it? If Fray Marcos was lying then he began to lie in a very big way from the beginning. He could not have intended merely to color the truth slightly, for not only did he make these statements positively about Cibola, he made them repeatedly, and there is no hint of that ponderous wording which tends to imply more than is actually said.

Before returning to Mexico to witness the consequences of this report it might be well to pause briefly and examine the strange reaction of the lord of Cibola to Estéban's decorated gourd. It came as a surprise because one would not normally expect a powerful ruler to be so merciless toward an unarmed group who clearly meant him no harm. We find a clue to the riddle in the word rendered "rattle" in the above translation; it is *cascabel* in the original, and it signifies a bell having a clapper trapped within an enclosure. The common sleigh bell is a good example, and because of the similarity in construction the same word is used to signify the rattles of a rattlesnake. Now what kind of cascabells did Estéban have on his gourd? If they were snake rattles then there is no obvious way to understand the Cibolan's violent reaction to them. But there were certain Aztec artisans in the vicinity of Mexico City who produced genuine cascabells of copper, silver, and also of gold by the sophisticated "lost wax" casting process [45]. Many examples of such bells have been found, even well into the present territory of the United States.

If these were the rattles which Estéban had attached to his calabash then they would have betrayed his origin. A few words of explanation from the messengers describing the foreigner who owned that gourd might easily have been enough to excite the Cibolan into a fury. For nearly twenty years had passed since the foreign conquerers had taken Mexico, and to be realistic one must imagine that the Cibolans were well aware of the fate of the Aztec nation. They might have looked upon Estéban, then, as the scout for an army that would later come against them as well, which, of course, he was.

Chapter 2:

Expedition of Conquest

WHEN THE FRIAR returned to Mexico with that glowing account of those marvelous cities it was only a matter of days before an army of volunteers began to assemble, eager to join an expedition of conquest. Francisco Vázquez de Coronado, the Governor of New Galicia, was appointed Captain General. He was a man of barely twenty-nine years at the time, but his meteoric rise was not due entirely to his skill as executive or strategist; other factors weighed heavily as well. Usually, as we shall see again, operations of this kind were privately financed, and the expedition to take Cibola was no exception. It was a joint venture, paid for partly by the Viceroy out of his own pocket and partly by Coronado himself—and perhaps also by his wife who was wealthy in her own right. The backers expected their rewards, then, from the wealth that they hoped would be taken, and a certain portion was claimed by the King as well.

It was agreed that the volunteers should rendezvous on Shrove Tuesday (1540) at Compostella, the capital city of New Galicia, and then and there begin their march for Cibola. Fray Marcos, having been promoted to the rank of Father Provincial, was to go along—as chaplain at least, but whether as guide remains to be seen. On the appointed day Mendoza himself harangued the assembled troops to encourage them on their way, and then, with colors flying, they were off! In keeping with the festive atmosphere, the Viceroy accompanied the army for the first two days of march, and then he returned to Mexico City.

The little army was a motley troop consisting of some 330 Spaniards, 1000 friendly Indians, and a few Negros. They were equipped with about a thousand horses which carried the provisions, and they also took some livestock on the hoof to be used for food along the way. Three ships laden with additional supplies were dispatched up the western coast to support the effort if a rendezvous could be realized. Only a few of the Spaniards were professional soldiers; the majority were young men from elite families who had come to the New World in search of their fortunes and were not gainfully employed at the time.

This is how it happened that the group sprang into being so quickly; they needed only the word. A great many more wanted to go, but a strict limit was placed on the numbers of both Spaniards and Indians in order that the security of the new colony should not be compromised. When one considers the situation thoughtfully he must marvel that this small, utterly unproven force should have had the daring to undertake an expedition of this kind against cities as great as Fray Marcos had described. The Spaniards had superior weapons, to be sure, but their fire-power was sorely limited so they would have no hope whatever against that multitude if the Cibolans should resist valiantly.

The army rested for a time at Culiacán, the last Spanish outpost along the way, where they took advantage of the opportunity to replenish their supplies. But there Coronado modified the order of march somewhat. Instead of moving ahead as a unit he decided to proceed with a smaller, more manageable detachment of fifty horsemen, a small infantry, Fray Marcos, and a few Indians and Negros; the rest of the

Expedition of Conquest

troops were to follow along behind in two weeks. This arranged, the general and that advance portion of the army set out on the next stage of march, but did they follow the same route that Fray Marcos had explored previously?

With hardly any exceptions historians agree that Coronado did truly retrace that former course. And why would he not? Marcos was present to point the way, and the grim alternative would have been to strike out blindly through uncharted wilderness to blaze an entirely new trail to Cibola. However, history records that the prize he finally won answered not at all to the friar's description. Indeed, the Cities of Cibola were found to be but modest little villages hardly different from the many others that dotted the region. Small wonder it is, then, that "our client" imagines herself to be the victim of an elaborate hoax. But to what purpose?

Because of these conflicting testimonies let us not be too quick to accept those routes as being the same. This a question that will have to be resolved by paying careful attention to the daitals. In either case we should expect to account for the discrepancies. Accordingly, we consult the other prime source of information about the route to Cibola, namely the letter that Coronado himself wrote to the Viceroy on August 3, 1540 after those small villages had been taken. He began that report by describing the journey northwards from Culiacán in these words [13;p.280]:

" On the 22nd of April last, I set out from the province of Culiacán with a part of the army. Judging by the outcome, it was fortunate that I did not take the whole of the army with me on this undertaking. The labors have been so very great and the lack of food such that I do not believe this undertaking could have been completed before the end of this year, and not without a great loss of life.

" Thirty leagues before reaching the place of which the father provincial, Fray Marcos, spoke so well in his report—the valley into which Fray Marcos did not dare enter—I sent

Melchior Diaz forward with fifteen horsemen, ordering him to make but one day's journey out of two, so that he could examine everything there before I arrived. He traveled through some very rough mountains for four days, and did not find anything to live on, or people, or information about anything except two or three poor villages. From the people there he learned that there was nothing to be found in the country beyond except the mountains, which continued very rough, entirely uninhabited by people. The whole company felt disturbed at this, that a thing so much praised, and about which the father had said so many things, should be found so very different, and they began to think that all the rest would be of the same sort. ..."

He has only just begun, but something is already amiss for he found the territory "very different" from what Marcos described earlier. As his report continues the General makes frequent reference to a "Valley of Hearts", and that deserves a brief explanation in advance. Recall that some four years prior to this expedition the Cabeza de Vaca party had passed through that same mountainous area. At one village along the way the four were feasted lavishly and were offered the dried hearts of 600 deer so they called the place "Hearts" in memory of the occasion. The village was situated in the mountainous region, but shortly after quitting it they negotiated a pass and emerged into the coastal lowlands. Most modern authorities think that village was near the present town of Ures (see Figure 1) and that the four continued from there down along the valley of the Sonora River. Its actual location is of no real concern to us here, but it is interesting and significant that Coronado should have thought to locate himself in that valley.

For let us recall that Marcos made no mention of the Valley of Hearts in his account, and how would he have identified it in any case? Estéban might have pointed it out if they had actually passed through it, but of course he and Marcos parted company very early along the way so that would not have been

Expedition of Conquest

possible. Then how did Coronado identify the valley? One might guess that they met natives who remembered the four from that previous occasion, but note in what follows that the General states, "I reached the Valley of Hearts, at last, ...", suggesting that it was no mere chance encounter but a planned goal. This question poses a little mystery of its own, the solution to which should become obvious very quickly. But now let us return to that report and read a few more paragraphs.

" I reached the Valley of Hearts, at last, on the 26th of May, and rested there a number of days. Between Culiacán and this place I could sustain myself only by means of a large supply of corn bread, because I had to leave all the corn, as it was not yet ripe. In this Valley of Hearts, we found more people than in any part of the country we had left behind, and a large extent of tilled ground. There was no corn for food among them, but I heard that there was some in another valley called Sonora. As I did not wish to disturb them by force, I sent Melchior Diaz, with goods to exchange for it. A little corn was obtained by trading, which relieved the friendly Indians and some Spaniards.

" Ten or twelve of the horses had died of overwork by the time we reached this Valley of Hearts, because they were unable to stand the strain of carrying heavy burdens and eating so little. Some of our Negros and some of the Indians also died here, which was a great loss for the rest of the expedition. They told me that the Valley of Hearts is a long five days' journey from the western sea. I sent to summon Indians from the coast in order to learn about their condition, and while I was waiting for these, the horses rested. I stayed there four days, during which the Indians came from the sea ...

" I set out from Hearts and kept near the seacoast as well as I could judge. But I found myself continually farther off, so that when I reached Chichilticale I found I was fifteen days' journey from the sea, although the father provincial said it was only five leagues distant and that he had seen it. We all became very distrustful, and felt great anxiety and dismay to see that everything was the reverse of what he had told Your Lordship. The sea turns toward the west directly opposite the Hearts for ten or twelve leagues. There I learned that the ships had been seen which Your Lordship had sent in search of the port of Chichilticale, which Fray Marcos had said was on the thirty-fifth degree."

These few paragraphs tell us all we need to know for the present so let us note the significant points in turn. Recall that in Paragraph 5 of his account Fray Marcos stated that he was writing down the names of the villages on another piece of paper. That piece of paper did not survive the years, and the name Chichilticale does not appear in his narrative. But we know from what Coronado has just written that it was a village near the sea at a point were the coastline turns sharply to the west. Judging from this part of the description it must have been located near the site of the present city of Guaymas, but in that case it would have been close to 28 degrees, not 35 as Marcos reported.

This is a fairly large discrepancy in latitude, but the only alternative would be to look for this village somewhere near the northern end of the Gulf where the coast makes another sharp turn to the west. At about 31 degrees, that would be closer to the figure Marcos specified, but the terrain itself argues against this idea. For Fray Marcos stated that the town he visited had an expanse of green fields around it; namely, there was arable farm land and the water to sustain crops. We find both of these resources in the vicinity of Guaymas, but not at the northern end of the Gulf. In actual fact, the thirty-fifth parallel passes more than 200 miles north of the extreme end of the Gulf so the friar was greatly in error with respect to the latitude in any case. Let us recall that Marcos was reputed to be skilled at navigation so assuredly he knew how to determine latitude by observing the sun and the stars, but here we must simply conclude that he did not have instruments at hand for making the necessary measurements accurately.

Expedition of Conquest

The second thing to be learned from Coronado's report is that this was the place where that rendezvous with the supply ships was to take place. Evidently they planned to stay near the coast so they could pass through that village. This obvious conclusion is confirmed by Coronado himself, for he states explicitly that he tried to follow the coast but failed in the attempt and found himself continually further from the sea as he advanced along a northerly course.

And then, finally, we learn that although the General never reached the coast to rendezvous with the supply ships, he arrived at Chichilticale nevertheless—but it proved to be fifteen days' journey inland from the sea, not one day's march as Marcos had described it. What an implausible error this is! And it becomes even more implausible upon closer examination: Many years afterwards one of the soldiers in that army, Pedro de Castañeda by name, was encouraged to write his recollections of the expedition, and here is how that chronicler remembered the episode at Chichilticale [73;p.313]:

" When the general had passed through all the inhabited region to Chichilticale, where the desert begins, and saw that there was nothing good, he could not repress his sadness, notwithstanding the marvels that were promised further on. No one save the Indians who accompanied the negro had seen them* and already on many occasions they had been caught in lies. He was especially afflicted to find this Chichilticale, of which so much had been boasted, to be a single, ruined and roofless house, which at one time seemed to have been fortified. It was easy to see that this house, which was built of red earth, was the work of civilized people who had come from afar."

---

* *The implied accusation is that upon hearing of Estéban's fate, Fray Marcos had immediately turned and hastened homeward in fright, writing in his report what the Indians had told him about Cibola and inventing the rest.*

So the coastal village of Chichilticale which Coronado expected to find, at which the rendezvous with his supply ships was supposed to take place, turned out to be a single mud house, standing in ruins, fifteen days' journey inland from the sea! What followed is anticlimactic, but let us continue along and learn how the expedition ended nevertheless. Leaving the old mud house behind the army pressed northward, and after fifteen days they spied "Cibola". Here is how Castañeda described that discovery:

" On the following day, in good order, we entered the inhabited country. Cibola was the first village we discovered; on beholding it the army broke forth with maledictions on Friar Marcos de Niza. God grant that he may feel none of them.
" Cibola is built on a rock; this village is so small that, in truth, there are many farms in New Spain that make a better appearance. It may contain two hundred warriors. ..."

Here was bitter disappointment indeed, a sorry prize to have been gained at such cost. But since those little pueblos differed in every possible particular from the description Fray Marcos had already sworn to one might ask why he consented to that identification. Did the inhabitants themselves call their villages Cibola? Coronado answers this question himself as he continues with his report.

" It now remains to tell about this city and kingdom and province of which the father provincial gave Your Lordship an account. In brief, I can assure you that he has not told the truth in a single thing that he said, except the name of the city and the large stone houses. Although they are not decorated with turquoises nor made of lime or of good bricks, nevertheless they are very good houses with three and four and five stories, very good apartments and rooms with corridors. There are some very good rooms under ground and paved, which are made for winter, and are something like hot baths. The ladders which

Expedition of Conquest

they have for their houses are all moveable and portable.
"   The Seven Cities are seven little villages, all having the kind of houses I have described. They are all within a radius of five leagues. Each has its own name and no single one is called Cibola, but altogether are called Cibola. ..."

And thus ends the saga of the Seven Cities of Cibola as history has preserved it these four and half centuries. It is a tale marred by such obvious inconsistency that one must be watchful if he is not to be deceived. Firstly, then, it might be well to ask of those little villages: By whom were they called Cibola? Not by the natives, surely, for they called themselves Zuni, and they continue to do so to this very day.

Many generations of historians have taken Coronado at his word and have judged the friar's detailed narrative to be a tissue of lies. Cowardice is given as the motive. Upon hearing of Estéban's fate Marcos supposedly feared for his own life, gave up all thought of proceeding to Cibola, and turned his feet homeward without further delay. But how could anything be more absurd? How could fear of proceeding to Cibola have prompted him to lie about the route that he followed in getting there? How does that explain calling an abandoned mud house fifteen days' journey inland from the sea a coastal town with a thriving population? All we know of Fray Marcos de Niza we have gleaned from reading his report. Questions of truthfulness aside, it is a sober, well composed and orderly narrative describing sound, reasoned behavior throughout; this is not the writing of a simpleton. If he were intent on lying he was certainly bright enough to have devised a story that would not inevitably be disproved by subsequent events. Coward and knave he might have been, but let us be wary of thinking him a fool.

Now it is very clear that in his report Marcos described a route only slightly inland from the sea. Indeed, when he had word at one point that the coastline turned sharply to the west he went to observe this unexpected fact for himself. On the other hand the army stayed well inland from the sea, and remarkably enough Coronado made little of this obvious discrepency. In particular, one might have expected him to question whether the friar had led him along an entirely different route than before, but this obvious possibility seems never to have entered his head; he charged him only with misrepresenting details in the terrain along the route. But since Coronado never examined those coastal regions his charge has no visible foundation so why should our critical attention not be turned to the General himself instead?

One can easily imagine that the army, burdened as it was with armament and pack horses, simply might not have been able to proceed along the exact route Marcos pointed out. Coronado may have been obliged to seek a detour, and, as things worked out, he continually found himself cut off from the coast by impassable mountains. This seems reasonable enough, but such factors could have been openly stated in his report to the Viceroy; there would have been no call whatever to impugn the priest's word or suggest that he was at fault in any way. As an alternative, one might suppose that Coronado recklessly disregarded his guide's direction, and in his impatience sought a more direct route northward. Later on, fearing the Viceroy's censure because of the outcome, perhaps he decided to claim that Marcos had actually led him along that inland course. It would be his word against the friar's.

But as one examines his letter to the Viceroy again with this possibility in mind he might easily conclude that Coronado had already decided to forsake the coastal route when he left that last outpost at Culiacán. One could then understand why he divided the army into those two groups; presumably he was preparing for rough going even though the friar had spoken of easy terrain ahead in his narrative. On this score alone one has reason to suspect that Coronado himself led the way northward out of Culiacán—and along a deliberately alternate route, for

Expedition of Conquest

according to Sauer [66;p.10] there are no more than isolated hills in the region south of Guaymas along the coast. They probably could have followed that route as a unit without difficulty just as the narrative had promised.

Although doubtless unintentionally, he offered support for this view in his own report, for recall the statement: "I ... kept near the seacoast as well as I could judge. But I found myself continually farther off, ..." He gave no hint here that he was yielding to another's direction; in fact, he could hardly have stated more clearly that the judgement of route was his own.

So one can find at least some support for that unsavory idea, but in truth this scenario is no more plausible than the other. On the one hand we have to picture a cowardly and lying priest who betrayed his trust and led the expedition astray into unexplored wilderness. And on the other hand we must imagine that the general of an army, a man who cherished his honor above all else, lied to his commander in order to place the blame for his own blunders upon the shoulders of another—and a man of God at that! However questionable may have been his motives for the expedition, such behavior would have been unrealistically petty and sacrilegious at that. Surely we must look elsewhere for understanding.

And perhaps we can find it by recalling that the expedition to Cibola was organized very quickly. Stimulated by a frantic lust for great riches it mushroomed into being almost overnight and became an irresistable tide that fed upon itself. Coronado himself was overcome with the fever of anticipation, and he concentrated all of his energies upon the growing enterprise. There was much work to be done so there was no time to lose if they were to be ready to leave by spring.

Accordingly, it was not until the expedition was well underway, perhaps during their journey between Compostella and Culiacán, that Coronado and Marcos would have had a chance to speak soberly and at length about those great cities that they had set out to conquer. The friar would then have described in detail their unending expanse, their countless warriors, and the vast, barren desert that surrounded and protected them. That would have given Coronado his first sound basis for realistically assessing the outcome of their venture, and he must then have realized that his little army was marching to its certain doom! Presumably Marcos had anticipated that a force worthy of the objective would have been recruited, but when he observed the small group that had actually assembled he saw clearly the futility of the enterprise, and he must have persuaded Coronado of that grim reality.

But what was to be done about it? The general could hardly turn back at that point; he would have been branded for life. His sense of honor required that he press forward—but he need not have pressed forward along the correct road! Perhaps he reasoned that it would be better to explore an entirely new territory, hoping to find wealth as yet unknown, than to persist along the road to certain death for all. But the fact that he accused Marcos of lying instead of merely being lost *requires that they were partners in the plan.* As a gentleman, Coronado would never have devised such a tactic, and neither could he have been so foolish as to expect his charges to stand if the friar would have chosen to defend himself. The obvious resolution to this seemingly incongruous situation, then, is that Marcos *offered* to play the role of liar so the army would not need to go against Cibola and be destroyed. It would be one man's reputation in return for the lives of thirteen hundred.

Presumably the army marched into the wilderness with the friar pretending to lead the way, but both he and the general knew full well that they would never arrive at their avowed destination. It must have been Marcos after all who identified that old mud house as Chichilticale and the Zuni pueblos as Cibola. So the report was written, the charges were made, and Fray Marcos de Niza was dispatched back to Mexico in disgrace along with the messenger who carried the report.

Expedition of Conquest

He would not dispute the accusations; on the contrary, he would meekly confess to them and spend the rest of his days in ignomy—an object of reproach to his brethren. One can easily imagine that their parting on that sad occasion would have been a sober and moving experience for both men.

So here at last is a realistic interpretation of the events that not only squares with the records, it also accounts for the discrepancies between them. In doing so it turns a seemingly farcical comedy of errors into a true-to-life drama that casts men in roles they could have played in good conscience. They were rational human beings, true to their creeds, who behaved responsibly at every turn; judged by the lights of their time, neither bears any stigma whatever. Did Coronado betray a trust by departing from the prearranged course? As commander in the field, having the more accurate information, his was the final responsibility, and he exercised it prudently. There was at least hope of reward somewhere along that alternate route, but there was none at all at Cibola. It was the only rational course, but had he followed it openly he could never have defended himself against the charge of cowardice that would inevitably have been leveled by his detractors, and he would not risk that calumny whatever the cost.

Likewise we now see Fray Marcos in an honorable role that is accurately true to life. Perhaps it is not easy to sympathize with his desire to propagate his faith through armed conquest, but that was the custom of the day. However, when the cause was doomed to failure he must have felt the blood of those 1300 on his own hands since he had been instrumental in bringing the effort about. His offering to be the lamb was therefore realistic, honorable and truly heroic. One can hardly imagine the strength of will required for a man of his caliber to play the coward and fool before the taunts of lesser men—and do it believably.

Where, then, was Cibola? If we wish to find those seven cities then we must ignore all the analyses that have gone before and take our lead from the one man who was there and returned to tell of it. Much scholarly argument would have been avoided if Fray Marcos had been more explicit with some of the details, but his narrative defines the way very well nevertheless. Only two legs of the trek remain after that singular point where the coastline turns sharply to the west so we can hardy go far astray as we join him there in Paragraph 12. In the next paragraph Marcos states that he was in "that valley for five days", and here the first problem arises because he did not say how far he was from that starting point on the coast when he made the statement. We have to guess.

It seems most probable that he would have entered the note in his diary just as he was leaving "that Valley", and if so then, since only five days are in question altogether, we cannot be far wrong if we guess that he was two days' travel from that notable point of the coast at the time. In all of Paragraphs 13, 14, and 15 he says nothing to indicate any further progress, but then in Paragraph 16, he records the passage of three more days and completes the first of those two remaining legs of the journey. Having thus traveled about five days since leaving that point near the sea where the coast turns sharply to the west he would have gained some 80 miles, and that would bring him near the site of the present city of Hermosillo on the Sonora River.

Now it is true that we had to guess at those two days, but that can hardly raise serious doubt about his position at this late stage of the journey. For reassurance one can lean upon a principle stressed by Carl Sauer in his attempts to analyze the Fray Marcos narrative [70]. Sauer urged, in brief, that the cities of today were the villages of yesterday, and the roads of today were their trails. This seems reasonable enough since population centers develop where the land is good and where water is available for agriculture. The most obvious conclusion one can draw is that the village which Marcos just left behind grew into the present city of Guaymas. Hermosillo is the next

Expedition of Conquest

main stop along the way northward today, and it fits, being about 80 miles distant.

One can hardly be certain of the exact spot, of course, but there is no need to be precise here because the target destination is so very great. The important point is that twelve miles away began an unpopulated region that required fifteen days to cross, and that could only have been the vast Sonoran Desert! One can appreciate the futility of the Coronado enterprise all the more at this point, for what would those thousand horses and thirteen hundred men have done for food and drink in that destitute wasteland?

Now Baldwin's translation misses a shade of meaning with respect to those fifteen days mentioned in Paragraph 17, a point which most other translators bring out explicitly. The original reads "... *hay <u>largos</u> quince dias de camino.*", that is, a *long* fifteen days' march. Although Fray Marcos said that they planned to cross the unpopulated region in fifteen days, apparently more than the normal exertion would be required in order to do so. Since the terrain at hand is not especially mountainous, then presumably more than the customary 15.5 miles would be covered each day. In order to gain some idea of what this might mean, let us note how Cleve Hallenbeck analyzed the problem. One ought to be aware that this author shared the common view that Marcos traveled through the same rough country that Coronado later described. Since the terrain in the Sonoran Desert is generally not so severe, Hallenbeck's conlusions might be regarded as safely conservative [40;p.43]:

> "   Marcos made his journey on foot over unimproved Indian trails, and tramping the old trails of the southwest is real work. I have covered hundreds of miles of them, afoot and mounted. ... By the time one has traveled fifteen miles under such conditions he has done a fair day's work, although he may have covered an air-line distance of no more than a dozen miles. ..."

Hallenbeck goes on to describe the many difficulties that would have confronted those travelers along the trail, and then he concludes as follows:

> "... I know from my own observations and experience that for a tolerable pedestrian on a march of a week or more on the trails of the semiarid Southwest, even today an average of sixteen miles a day is fair going, twenty miles a very good pace, and twenty-four miles an exceptionally fast rate that I doubt could be maintained for a week under summertime conditions."

The air-line distance gained must be somewhat less than the distance covered on the ground, but the difference in our case need not have been great because the terrain in the Sonoran Desert is not generally as rough as Hallenbeck has considered here. We must also keep in mind that Marcos followed a well-worn path, one which had been determined by years of Indian travel to be the best possible route. Therefore, since Fray Marcos and the Indians were obviously "tolerable pedestrians", well hardened to the trail, let us grant them the twenty miles per day that Hallenbeck says is a very good pace. In that case they should have covered some 300 miles in those fifteen long days.

Figure 2 illustrates the conclusions reached thus far. The friar's approximate position in Paragraph 12, the site of the present city of Guaymas is indicated by the letter "G". His position five days later is given by the "H", corresponding to Hermosillo, and an arc of 300 miles radius has been swung about this point as center. Presumably his destination was not far from this curve. The cities could hardly have been situated much to the east or the Coronado expedition would have spied them in passing. Neither could they have been located further to the west, closer to the mouth of the Colorado River, since this region was examined by other parties during that same time period, and nothing was found. So there can be little doubt that

FIGURE 2

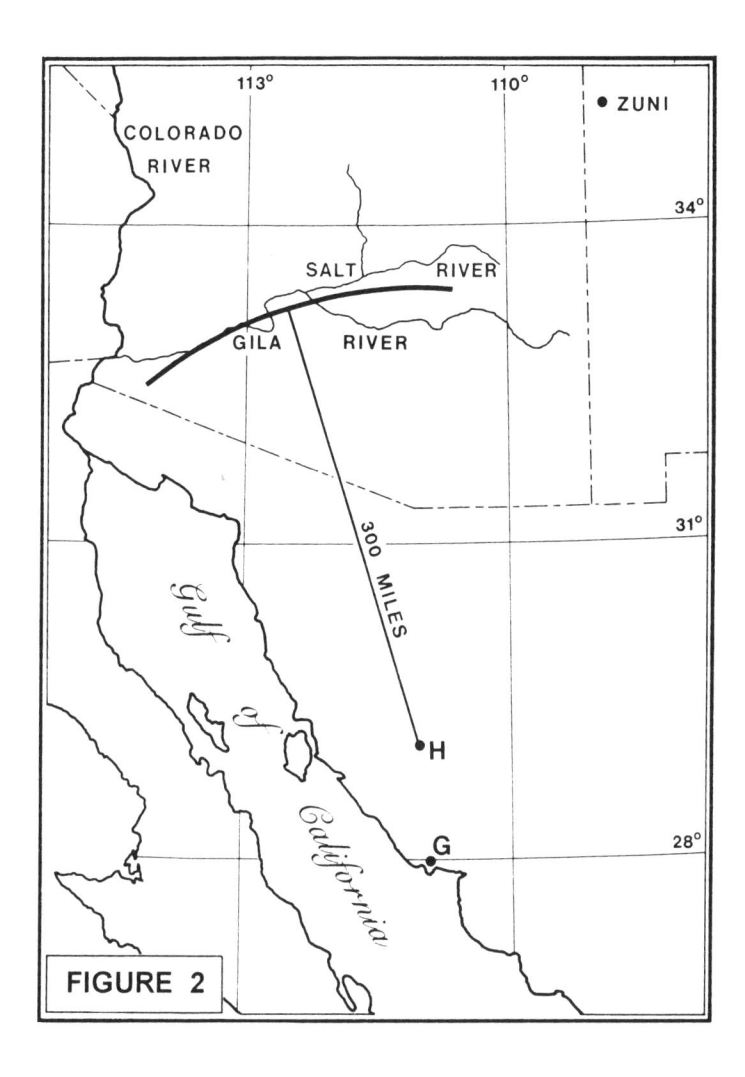

if the great cities ever existed at all then they must have been situated somewhere within that central region. This is an encouraging conclusion indeed because either the Gila River or the Salt River would provide enough water to support a substantial population.

Fray Marcos de Niza played his part faithfully to the end.

He went to his grave on the 25th of March in 1558 without uttering another word about the matter as far as anyone knows, and he has been known to historians as the "Lying Monk" ever since. On that day the curtain fell on the first act of this great drama. The obvious question now is this: If Fray Marcos was the first, then who was the next European to enter that region, and what did he find there? Oddly enough, the *entr'acte* was a long one for we must wait a great many years to meet the next man with a pen who came anywhere near.

Chapter 3:

THE SEQUEL

AFTER THE ZUNI villages had been taken, Coronado cast about to see if the new land held anything of value that might redeem the expedition. He sent scouting parties east into the panhandle region of Texas and also westward as far as the Grand Canyon of the Colorado, but they found nothing of interest anywhere. Hearing reports of a rich kingdom to the northeast the General himself led a small detachment well into what is now Kansas only to find that he had been deceived by a deliberate hoax; it was the ruse of a wily Indian who tried to lure the army to its destruction. Theirs was a valiant but disappointing effort. They searched an enormous territory, but they found no riches whatever.

Then, early in the spring of 1542, Coronado suffered a serious head injury during practice maneuvers, and his life hung in the balance for a time. He regained his feet at last, but that former driving ambition was gone; the country was destitute and worthless, and he yearned only to return to Mexico and home. But many of the troops were strongly opposed to abandoning the newly gained territory despite its poverty so they began the homeward trek a divided and demoralized band, their spirit and discipline badly decayed. When they had set out from Mexico two years before every man among them had been the envy of all who were left behind. Without a doubt they would each gain wealth, lands and titles and return to tell glorious tales of conquest. As it turned out they straggled home a beggarly lot with the General himself on a stretcher near the end—but they did return. Despite the Viceroy's disappointment Coronado was allowed to resume his duties as Governor of New Galicia, but he was removed afer two years when his ability to manage the office became doubtful; apparently he did not fully recover from his injury. However he continued to serve as a minor public official for another decade, and he died peacefully in 1554 during his 44th year.

The dismal reports from that expedition effectively shattered the dream of riches in the north. Additional tales of Cibola that might have filtered south could only have been met with snickers of derision since no further efforts were ever made in that direction. If Coronado had reported merely that he searched the region and found no such cities then interest might have been rekindled by new and specific accounts, but, as it was, Cibola had been found and identified. The matter was settled. And such tales would likely have become less and less frequent as time went by because of the massacre of Estéban and his party. Those Indians who lived directly on the road to the great cities would be afraid to go back, as they themselves had said, and they would pass the word along to others who were going there as well. Moreover, the Cibolans themselves might have thought it prudent to bring that former traffic with the south to an end; if so they could have devised still other methods for discouraging it. In view of the circumstances, then, perhaps it is not surprising that the Seven Cities remained as they were, veiled in obscurity, while the years turned slowly into decades.

And the decades passed freely for we must wait a century and a half to meet the next European who ventured into that region. He was Eusebio Francisco Kino, a Jesuit priest and missionary originally from the Tyrolean district of Austria.

The Sequel

Kino had no particular interest in missionary work until, as he believed, he was miraculously cured of a serious illness. After his recovery he volunteered for foreign service and requested assignment to the most dangerous of the mission fields—his first choice being the Mariannas, or, failing that, China. But he was ordered to Mexico instead, to which he embarked early in the year 1681 out of Cadiz.

Father Kino was a man of many interests, including both astronomy and mathematics, and upon arriving in the New World his first act was to see to the printing of a short monograph detailing his theories and observations of the great comet that had only recently receded. This done, he pursued his assignment to the San Bruno settlement in (Lower) California were he remained for six years. But eventually that colony had to be abandoned so he returned to Mexico to receive his next assignment as missionary to the Pimas. Their domain, Pimería Alta as it was called, comprised most of what is now Sonora and that part of Arizona which lies south of the Gila River. It was bounded on the east by the land of the Apaches and on the west by the Colorado River and the Gulf of California.

Kino arrived at the frontier of settlement in 1687 and founded his first mission, *Nuestra Senora de los Dolores*, at a site near the present town of Magdalena, about a hundred miles north of Hermosillo. This would be his home and headquarters for the remainder of his life, but he explored extensively to the north and northwest and established a string of missions that extended as far north as Tucson. Let us hear Bolton summarize Father Kino's accomplishments because he confirms the fact that this priest was the first white man to enter that region since the time of Fray Marcos [15; p. 53]:

" Kino's work as missionary was paralleled by his achievement as explorer, and to him is due the credit for the first mapping of Pimería Alta on the basis of actual exploration. The region had been entered by Fray Marcos, by Melchior Diaz, and by the main Coronado party, in the period 1539-1541. But these explorers had only passed along its eastern and western borders; for it is no longer believed that they went down the Santa Cruz. Not since that day—a century and a half before—had Arizona been entered from the south by a single recorded expedition, while, so far as we know, not since 1605, when Oñate went from Moqui down the Colorado of the West, had any white man seen the Gila River. The rediscovery, therefore, and the first interior exploration of Pimería Alta was the work of Father Kino."

During his journeys northward from the Mission Dolores, Kino continued to hear reports of deserted cities further to the north, but it was not until 1694 that he actually mounted an expedition and set out to investigate them. Here is how he described that incident in his memoirs [15;p.127];

" In November, 1694, I went inland with my servants and some justices of this Pimería, as far as the *casa grande*, as the Pimas call it, which is on the large river ... that flows out of Nuevo Mexico and has its source near Acoma. ...

" The *casa grande* is a four-story building, as large as a castle and equal to the largest church in these lands of Sonora. It is said that the ancestors of Montezuma deserted and depopulated it, and, beset by the neighboring Apaches, left for the east or *Casas Grandes**, and that from there they turned towards the south and southwest, finally founding the great city and court of Mexico. Close to this *casa grande* there are thirteen smaller houses, somewhat more dilapidated, and the ruins of many others, which make it evident that there had been a city here. On this occasion and on later ones I have learned and heard, and at times have seen, that further to the east, north and west there are seven or eight more of these large old houses and the ruins of whole cities, with many broken metates and jars, charcoal, etc. These certainly must be the

---

* Obviously another site entirely, presumably located somewhere in northern Mexico.

The Sequel

Seven Cities mentioned by the holy man, Fray Marcos de Niza, who in his long pilgrimage came clear to the Bacapa* rancheria of these coasts, which is about sixty leagues southwest from this *casa grande,* and about twenty leagues from the Sea of California. The guides or interpreters must have given his Reverence the information which he has in his book concerning these Seven Cities, although certainly at that time, and for a long while before, they must have been deserted. ..."

Here is the first statement of a riddle that would be a stumbling block for many years because the spectacle that Father Kino gazed upon that day bore all the obvious marks of extreme antiquity. A century and a half had passed since Fray Marcos had observed those cities alive and thriving, but even that great span of time seemed wholly inadequate to account for the ruin that lay before him; Kino thought the cities must have been deserted even longer than that. And yet the picture was one of strange contradiction as closer inspection revealed. Three years later, in the year 1697, accompanied by a troop of soldiers, Father Kino made another expedition into the north, and he stopped again at this *casa grande.* He recorded the following additional details on that occasion [15;p.172]:

" The soldiers were much delighted to see the Casa Grande. We marveled at seeing that it was about a league from the river and without water; but afterward we saw that it had a large aqueduct with a very great embankment, which must have been three *varas*† high and six or seven wide—wider than the causeway of Guadalupe at Mexico. This very great aqueduct, as

---

* *The letters B and V are all but interchangebable in the Spanish language so the names were probably the same; Father Kino merely spelled the word differently. But this must be a coincidence since Marcos passed through Vacapa very early in his trek (Paragraph 3); the site mentioned here by Kino is therefore far to the north.*
† *A vara is just slightly less than 33 inches.*

is still seen, not only conducted the water from the river to the Casa Grande, but at the same time, making a great turn, it watered and enclosed a champaign many leagues in length and breadth, and of very level and very rich land. With ease ... one could now restore and roof the house and repair the great aqueduct for a very good pueblo, ..."

So the great house itself was not in such a bad state of repair; at that time it would have been easy to restore the building and make it serviceable again. And even more definite testimony can be cited to this same effect, for among the soldiers escourting Father Kino on that journey was a Lieutenant Juan Mateo Manje who kept a detailed diary of his experiences. Some years later Manje wrote a book in which he told of his adventures in this new land, and there he described the events at hand as follows [51;p.85]:

" We continued to the west. After four leagues, we arrived at mid-day at *Casas Grandes,* inside of which Father Kino said mass even though he had traveled without eating until then. One of the houses was a large building four stories high with the main room in the center, with walls two *varas* in width made of strong *argamasa y barro* and so smooth inside that they looked like brushed wood and so polished that they shone like Puebla earthenware. The corners of the windows are square and very straight, without sills or wooden frames. They must have been made in a mould. The same may be said of the doors. ..."

The word picture that he paints here calls to mind a very well preserved structure indeed; there is no suggestion whatever of erosion or decay even though the city at large lay utterly in ruins. In fact, the scene appears so incongruous that one might suspect that he overstated the case somewhat in order to heighten the sense of drama in his book. However, there still exists today an original manuscript, signed by Manje himself,

The Sequel

PLATE 1: *The Casa Grande ruin as it appeared around 1900. (Reproduced from an undated print in the Arizona Historical Foundation collection at the Arizona State University Library).*

in which he recounted those same events again in slightly different words. A translation of this other document is included as a supplement in Reference 51, and the portion of interest reads as follows (p.287):

> " ... The walls of a strong conglomerate of dried mud, two *varas* thick, were so smooth and polished that there was not the slightest hole. Likewise the corners of the windows and doors were so straight and regular in size that they looked like they had been moulded and were smoothed to a fine finish. The natives of this vicinity had set fire to and burned the roofs, which were of a non-decaying timber. ..."

Evidently, then, although the smaller buildings making up the rest of the city lay in ruins, this great structure was in surprisingly good condition at the time. The witness stressed this point particularly, and he said it not once but twice. Now it is important to notice that the building was made of mud so even the rain would erode it, as is evident from the photo in Plate 1 which shows the Casa Grande as it appeared at some (unspecified) time near the turn of this century, about 200 years after Manje saw it; there can be no mistaking the changes which the wind and the rain have wrought. Obviously this great degradation could not have progressed far by Manje's time or he would not have felt moved to marvel at the smooth, flawless walls and the straight doors amd windows.

It is true that Manje spoke especially of the inside walls in the passage quoted from his book, but he also mentioned the doors and windows as being in similarly very good condition. And, since the roof and been burned, even the inside walls were exposed to the weather by the time he saw them. Moreover the roof beams seemed new, as if they had been hewn from non-decaying timbers. In order for the building to have been in such fine condition, then, one would think that it must have been in use and actively maintained until only a very short time before the Lieutenant came upon the scene—a time short, that is,

The Sequel

PLATE 2: *The Casa Grande as it remains today, a National Monument near Coolidge, Arizona.*

compared to two hundred years.

The Casa Grande is now preserved as a National Monument, and Plate 2 shows the structure as it is today. Some of the damage has been repaired around the base, and a roof has been erected to protect it from further damage from the rain; nothing whatever remains of the rest of the ancient city.

Continuing with his account of that expedition into the unknown lands, Manje goes on to describe an interesting site just downstream fom this Casa Grande. His narrative continues as follows [51;p.87]:

"   On the banks of the river at a league's distance from *Casas Grandes* we found a settlement ... where we counted 150 souls to whom we preached eternal salvation. The priest baptized nine children. they had fear of the soldiers and horses since they had not see them until this time.

"   On the 19th [the next day], after mass, we continued to the west over arid plains. On all lands where these buildings are located there is no pasture. It seems that the land has a saline character. After having traveled four leagues, we arrived at a settlement called Tucsoni Moo, named thus on account of a great mound of wild sheep horns piled up, looking like a mountain. These animals are so plentiful that they are the people's common source of sustenance. This pile of horns is so high that it is higher than some of their houses. It appears as if there are more than 100,000 horns. The heathen Indians welcomed us profusely, sharing with the soldiers some their supplies. We counted 200 courteous and peaceful people. ..."

We recall from Paragraph 10 of the Friar's narrative that the people of Totonteac wore clothes of wool which they obtained from animals the size of Castilian greyhounds. Surely here is the source of that wool. The great number of those sheep suggests that they had been domesticated by the former inhabitants and preserved as a resource. That they were still plentiful at the time (1697) further suggests that they had been freely slaughtered for only a relatively short time. On the basis of this great herd of sheep, then, we might identify the ancient province that surrounded the Casa Grande as Totonteac.

The central regions of Arizona continued wild and raw for a great many years after the passing of Father Kino in 1711. As a consequence of the Mexican War of the 1840's and the Gadsden Purchase of 1853 the region came under the jurisdiction of the United States, and settlers began to arrive. However, because of hostile Apaches, the colonies did not prosper until the end of the Civil War when the Army moved in to

The Sequel

enforce peace upon the region. In 1867 Jack Swilling noted the remains of ancient canals leaving the Salt River, and seeing the possibilities, he decided to undertake farming by irrigating lands remote from the river itself. He acquired a modest backing, and the Swilling Irrigating and Canal Company was formed to deliver the stuff of life again to the parched land.

Soon small farms dotted the newly irrigated area, and since barley and pumpkins were the chief crops at first the budding new town came to be called Pumpkinville. But eventually the 300 residents decided that their little community deserved a more dignified name so a town meeting was called for the purpose of selecting one. Darrel Duppa, an Englishman of considerable background and learning, is usually credited with the suggestion that was finally adopted. He recounted to the assembly the ancient legend of the Phoenix bird; it lived for five hundred years, was consumed by fire, and was then reborn anew from its own ashes. Duppa felt that Phoenix would be an appropriate name since, as was plainly evident to all, the new town was rising from the ashes of an ancient civilization. The residents liked the idea, so Phoenix it was.

The first methodical excavation of the ruins around and about the area was undertaken by Frank Hamilton Cushing in the years 1887 and 1888. The project was known as the Hemenway Southwestern Archaeological Expedition because it was financed by a Mrs. Mary Hemenway of Boston. Cushing set up his camp at a location about six miles south of the river and to the east of a range of rocky hills now known as the South Mountains. He called the site "Los Muertos"—that is, "The Dead". After three seasons of work at this one location the expedition shipped three railroad cars full of excellent material back to the Peabody Museum of Harvard, but more than 50 years passed before a report of the findings was published. By that time Cushing was long since dead so one must search elsewhere to learn of the opinions he formed from his own on-site observations. We shall return to consider them later.

New Insights to Antiquity

While the Cushing party made careful note of the Los Muertos site and the ancient canals serving that region south of the Salt River, the remainder of the valley was not so carefully studied. Although some of the old canals were preserved and put to use almost as they were found, and others were used in part, most of them were filled in and new ones were constructed. Consequently not much remained of the old system after a few years; and where the ancient cities intruded they also were put to the plough, usually with few records being kept. It is a great tragedy that this destruction continued for some considerable time before anyone thought to make a careful record of the ancient system as a whole. In fact, it was not until 1903 that James W. Benham [10] produced the first detailed map that attempted to trace the courses of all the canals in the old system. That map is reproduced on the end papers of this book*. The scale is given by the large squares (township boundaries) which are six miles on a side. It is important to keep in mind that the system was far from intact when this map was drawn, although Benham said that he based his work on observations that had extended over many years. Companion notes were written by H. R. Patrick [65], and the combination of map and notes was published as Bulletin No. 1 of the Phoenix Free Museum. Patrick himself had been a resident of the region for twenty-five years, and he also had studied those remains as best he could so we can hardly do better than to hear him at length.

" The size and capacity of the canals are quite surprising, the largest being seventy-five feet wide between the centers of the borders and probably not less that forty feet wide in the

---

* Let it be noted that the map has been edited slightly in order to use the space at hand most effectively. To this end approximately five eighths of an inch have been cropped from both the left and right edges. The legend has been preserved intact by moving it to the right appropriately.

The Sequel

bottom of water under way, with borders about six feet in height being quite equal to any canal of the present system.

" The longest canal is about twelve miles in length but one of the old systems has about twenty-eight miles of mains, while in the aggregate there are one hundred and thirty-five miles of main in the old system. While the total mileage of the modern system is but ten miles more.

" The acreage of land under those old systems is approximately one hundred and forty thousand acres, which, if divided into small holdings such as the present Indians cultivated under their natural conditions must have represented over twenty thousand farms, and with a corresponding number of persons to each family, the ancient canal system must have supported a population of from 120,00 to 130,000 people,—but to this may be added a large population in the cities who may not have been farmers or tillers of the soil, so that the population of the entire valley might easily have been 200,000.

" Coming to the subject of cities and towns, we find them covering an extent of country about twenty five miles in length, east and west, and fifteen miles in width north and south.

" The more important of these cities are seven in number, and are designated by the letters A to G on the map, and are noted for having one large principal building, probably a communal house or temple, around which are clustered from one hundred to two hundred small buildings, besides these there are seven or eight small towns or villages that must have contained a numerous population, and there are other isolated ruins that are scattered along or near the ancient canals."

So there were seven principal cities of old supported by that vast irrigation system in the valley of the Salt River. They all derived their vital water from the same source so one can deduce that they must have been confederated under a single governor—otherwise they could not have lived in peace during times when the river was low. Can there be any doubt, then, that these were the seven renowned cities of Cibola?

It is sad to report that out of all those seven cities only a single, small residue of one solitary structure remains in evidence today. This was the principal building of the city denoted by "D" on Benham's map. The relic has been named "Pueblo Grande", and it is now preserved as an archaeological site and museum by the City of Phoenix. One corner of this old ruin is shown in Plate 3; it is here viewed from the south with the modern Grand Canal being visible in the foreground.

Phoenix is bounded on the south by a range of low, rocky hills called simply the South Mountains. In its wildest state it

PLATE 3: *Looking north at the Pueblo Grande with the modern Grand Canal in the foreground.*

The Sequel

was a barren, uninviting tract neither grand in its aspect nor pleasing in its verdure. Nevertheless, in the early 1920's a group of Phoenix residents were able to appreciate its potential as a park site, and they set themselves to the task of developing the region into a recreation area. South Mountain Park is today mostly a wilderness preserve comprising in excess of 15,000 acres, but paved roads have made it accessible to the public. Picnic accommodations have been provided, and a number of trails have been cleared for hiking and horseback riding.

In 1926 a curious memento was discovered on the eastern slope of the South Mountains—a rude inscription, dated 1539, with the name Fray Marcos de Niza clearly legible. The location was held secret for several years until the property

PLATE 4: *The Fray Marcos de Niza inscription on the eastern slope of South Mountain.*

« 63 »

could be annexed to the park and the inscription protected against vandalism by a closely spaced grating of heavy steel bars. A photograph of the inscription, taken through these bars with a wide-angle lens, is reproduced here as Plate 4. The damage visible at the right was inflicted before the barricade was erected; in fact, it may have occurred even while the writing was being made since it appears in the very earliest photographs.

Plate 5 shows the site with its protective enclosure from a few yards to the east, and one can clearly see the very dark color of the rocks that cover this hill. Even so, only a very thin

PLATE 5: *Showing the site of the Marcos de Niza inscription and the protective barricade of heavy steel bars.*

The Sequel

crust near the surface is dark; the interior is much lighter, which is why the writing is so clearly legible. It has not been artificially enhanced in any way.

As might be expected, most authorities have branded this inscription a counterfeit, and not a very clever one at that. For as everyone knows, the Coronado expedition passed far to the east and missed this spot by probably 150 miles. And of course, the presumption is that in so doing it retraced the route followed earlier by Fray Marcos. But not only had the supposed forger placed his work far away from the "correct" course, he had seemingly made an even more foolish mistake in his choice of text. In order to understand this problem, however, and to put it in its proper place, we must digress briefly and review a brief but tragic chapter out of the early history of New Mexico.

In the years following Coronado's expedition a Spanish settlement was established in that new territory; Santa Fe was its capital. But in the course of time resentment mounted amongst the Indian population and eventually an embittered native organized a revolt against the Spanish rule. A rope with a number of knots was circulated secretly around the various Indian villages and was duplicated at each one. One knot was to be untied each day, and when none remained the white men were to be destroyed. As this worked out the appointed day was August 13, 1680, but word of the plot reached the authorities a few days ahead of time.

When the insurgent learned of this breach he ordered the attack to begin at once, so hostilities actually started on August 9th. About 400 settlers were killed, along with the priests, and only those few who were warned in time managed to survive and to work their way to Santa Fe. The capital could not be held for lack of supplies, but there were enough Spaniards present to enforce an orderly retreat; they moved south, about to the present site of El Paso. This is how it happened that by the end of the year 1680 not one white man remained alive within all of northern or central New Mexico.

Twelve years passed before the Spaniards were able to launch an effective counterattack. As has already been mentioned, the Spanish government did not normally finance campaigns of this kind. Instead, a contract was let to some man of means to undertake the task. If no plunder was likely then his reward would be in lands, titles, or other honors, and this case was no exception. The expedition was under the command of Don Diego de Vargas who was appointed Governor and Captain General; he provided the necessary supplies and resources from his own personal fortune, which was extensive.

PLATE 6: *The de Vargas inscription at El Morro National Monument, New Mexico. National Park Service photograph.*

The Sequel

Now Burke [18;p.143 *et seq.*] has reported a great many details of that expedition. The General started north up the valley of the Rio Grande with a small band of troops in August of 1692, but he found all the pueblos along the way deserted, the inhabitants having fled. His first contact with Indians in this northern excursion was at Santa Fe where he arrived on the 13th of September. De vargas was a capable diplomat and resorted to the force of arms only as a last resort, so he was able to retake Santa Fe at the conference table, so to speak, without a shot being fired.

According to Burke, de Vargas left Santa Fe on the 21st of September bound for Pecos to the east. That pueblo retaken, also by persuasion, he returned to Santa Fe and then headed west. Sometime during this western campaign, before the end of that year, he caused the writing shown in Plate 6 to be inscribed upon a massive rock a few miles east of Zuni. The text is heavily abbreviated, but as presently understood it is rendered into plain Spanish as follows [7]:

> *Aqui estubo el General Don Diego*
> *de Vargas, quien Conquistó*
> *a nuestra Santa Fé y a la Real*
> *Corona todo el nuebo*
> *Mexico a su costa*
> *año de 1692*

The discerning reader will notice that, excepting only for the date itself, the last three lines are identical to the writing on the rock in South Mountain Park. In English this becomes:

> Here was General Don Diego
> de Vargas, who conquered
> for our Holy Faith and for the Royal
> Crown all the New
> Mexico at his own expense
> year of 1692

As his second blunder, then, the supposed forger had slavishly copied the last lines of the El Morro inscription, being unaware of their true meaning. For clearly, if these words have been correctly interpreted then they are utterly out of place upon a rock in South Mountain Park.

But one ought to grant any supposed counterfeiter more devotion to his art than this would imply so it behooves us to examine the above interpretation very carefully, and indeed, several serious flaws in this reading are readily apparent. For one thing, note that the small **e** above the first **a** in the third line of the inscription has been entirely ignored in rendering the text into clear Spanish. For another, the letters that have been interpreted as *todo* in the spanish (all, in English) form not one word in the original, but two, and the second word or abbreviation is capitalized. And finally, while it was true that de Vargas financed the expedition, it would have been out of character for a man of his position to have mentioned the fact on a monument of this kind. Furthermore, it was the custom in those days for the commander to bear the cost of such an effort so it was already understood by all and would not have been worth recording in this fashion.

But the first error is certainly the most serious and the root of all the difficulty because if the clear text is not recovered correctly then the meaning will surely be confused. The third line, then, must start with a word beginning in **a** and ending with **e**, and that word most probably is a preposition. Only one possiblity comes to mind; it the word *allende*, meaning beyond. In that case *Santa Fé*, which was rendered "Holy Faith" above, should not be translated at all; it ought to be read simply as the name of the capital city, but if so then the added word *nuestra* deserves some comment.

A plausible interpretation for this unexpected usage is suggested by an alternate meaning for the word *real*. Thus, as an adjective it signifies royal, but the word is sometimes used as a noun also, in which case it denotes the encampment of an

The Sequel

army*. In fact, the formal name of the city is *Villa Real de Santa Fé de San Francisco,* and it harks back to a former Santa Fe after which it was named. When Ferdinand and Isabella were in process of driving the Moors from Spain, the last Moorish stronghold to be besieged was at Granada. Here the soldiers constructed an uncommonly elaborate camp-city outside the city walls. So elaborate was this encampment, in fact, that they thought it deserved a name; it was called Santa Fe. It is interesting to note that Columbus proposed his voyage of discovery to the King and Queen while they were in residence at that first *Villa Real de Santa Fé.*

Then "Our Santa Fe" may simply have been meant to emphasize its connection to that first tent city and to distinguish it therefrom. The usual construction, "New Santa Fe", might have been deemed inappropriate in this case since that would tend to imply a new and different faith. With this understanding, then, the first three lines of the inscription become:

> Here was General Don Diego
> de Vargas, who conquered
> beyond Our Santa Fe and to the encampment

—the encampment, that is, of his own army. In that case these three lines are complete in themselves; the next line must begin a new thought altogether, and the very construction bears this out because in the Spanish language an adjective nearly always follows the noun it modifies. Accordingly, if *real* had been meant as an adjective describing crown then we should have expected the scribe to write *la corona real.* But he did not,

---

* According to present usage, real *is deemed to be of the masculine gender so it should take the article* el, *but we note that the feminine article,* la, *is used in the inscription. Therefore this interpretation assumes that the usage in those days was not universally fixed. Presumably there were local or dialectical differences in the article used with this noun.*

« 69 »

giving additional grounds for interpreting *real* as a noun even though the article does not conform to present standard usage.

With this minor discrepency, then, this interpretation not only agrees with the writing on the rock, it also agrees with history. That is, it does not ignore that superscript **e** above the first **a** in the third line, and, as has just been seen, de Vargas did go beyond Santa Fe to the east to retake Pecos before turning around and going west. On the other hand, the currently accepted reading is not historically accurate since he had not retaken "all the New Mexico" by the year 1692. The campaign was only three months old when those words were inscribed; he had only just begun. Furthermore, the prevailing interpretation requires an unlikey syntax in which an adjective precedes the noun it modifies as well as an unsupportable reading of the two groups, *to* and *Do*, as a single word.

The meaning of the next two lines is obscure, but if they constitute a new thought altogether then we shall be able to reach a satisfactory resolution to our present problem without understanding them. However, if the word *su* (which was interpreted as "his own" above) does indeed refer back to de Vargas then these two lines would belong to the previous thought, and all five lines would have to be interpreted together. But *su* could just as well be read as "her" or "its", and *costa* also has a double meaning—it being the common word for coast. However this word is applied in a broader sense in Spanish than in English, for let us recall Father Kino's usage in a passage quoted earlier:

> " ... These certainly must be the Seven Cities mentioned by the holy man, Fray Marcos De Niza, who in his long pilgrimage came clear to the Bacapa rancheria of these coasts, which is about sixty leagues southwest from this *casa grande*, and about twenty leagues from the Sea of California."

Since the Bacapa rancheria was so far inland from the sea (more than 60 miles) it would not be called a coastal

The Sequel

settlement as we would use the term, and moreover, he used the word in the plural. Evidently then, it is not the exact equal of its English counterpart. In that case, the fifth line might be interpreted:

> Mexico to her coast

Perhaps it was meant in the sense, "Mexico, homeward bound", but obviously the meaning could depend critically upon the sense of the previous line—which in turn hinges on the meaning of Do, at present unknown. But it is now surely plausible that these two lines are indeed independent of the first three, and in that case perhaps their import can be discerned even if their meaning is obscure. Namely, it is not impossible that the relic in South Mountain Park is indeed a forgery, but if we assume it to be genuine then it must be considered strange that these two inscriptions, separated in time by over a century and a half, should both contain two unusual lines, in the same order and identical even to the form of the letters.

Perhaps the first step toward resolving this little puzzle is to note that the first three lines in the de Vargas inscription contain hardly a word that is spelled out in full; those lines are characterized by their artful abbreviations. On the other hand the next two lines contain only one abbreviation, if we understand Do to be such. But notice especially that this one is executed altogether differently than the abbreviation of the name "Diego" in the first line. This variation in technique, coupled with the pronounced change in the manner of forming that one solitary abbreviation, suggests that the writer may there have been copying from another text.

Now de Vargas probably did not inscribe the writing upon the rock himself although he may have directed that it be done. Someone else executed it, and the very precise form of the letters suggests that the writer was highly literate and practiced in his penmanship—a priest without question. In fact, Burke [18;p.146] states that three priests accompanied de Vargas northward and that all were Franciscans.

A glimmer of light peeks through for we can be confident that in addition to the narrative that we have already read, which was an official government document, Fray Marcos would have filed a report of the journey with his own order as well. But it need not have been merely a copy of the other; in fact, we recall that Father Kino mentioned a "book" which Fray Marcos had written. His official report would hardly be called a book so the account written for his own order was presumably much more detailed. Perhaps he would even have reported the very words that he inscribed upon that rock above Cibola. Later Franciscans would have access to that book. Granting this much, who would say that those three Franciscans bound for that same country (so they thought) would have failed to read that document?

Accordingly, the most plausible explanation for the correspondence between the two monuments is not that a counterfeiter copied the de Vargas inscription, but that the de Vargas scribe deliberately quoted the previous inscription as Fray Marcos had reported it. Presumably the scribe intended this token as a quiet tribute to his maligned brother of the cloth. Although this residual question cannot be considered critical to our present study it is a matter of interest for its own sake. One would hope that on some future occasion the archives of the Franciscan order will yield up that second narrative which Marcos must have written. When it is found then the point will be settled but probably not before.

Chapter 4:

DEAD MEN

LET US RECALL that Father Kino thought that the former inhabitants of the Casa Grande had moved southward, having been weakened by prolonged wars with the Apaches. The various Indian villagers living in the area around about offered differing opinions, but apparently they simply did not know who those people had been or what had become of them. The more recent inhabitants called them simply "Hohokam", and that has come to be the accepted name of that ancient nation today.

The most widely popular interpretation of this Pima word would have it signify "Those who have gone", and thus bring to mind a migration, or an exodus of the former inhabitants. But this is not the proper meaning of the word at all as Haury [43;p.5] has taken pains to explain in detail. This author points out that Hokam is the Pima word for anything whatever that is "all used up", and repetition of the first syllable forms the plural. Thus, the term could be used to describe the houses or the cities perfectly well since, lying in ruins, they were all used up. But the name is not applied to the cities; it is given to the people who occupied them. According to Turney [79;p.21], Hokam is the Pima word for a dead man, and this specific usage is perfectly consistent with the general meaning because a dead man is surely all used up. Hohokam, then, signifies "Dead men". Perhaps the name was applied first to the cities and only afterwards to the people who had occupied them. But if it was given to the population from the beginning then presumably when first found the ruins were strewn with skeletons of the former inhabitants!

The cities which formerly occupied the valley of the Salt River, then, have been identified as the Cities of Cibola, while the region surrrounding the Casa Grande fits well with the wool-growing province of Totonteac "to the southeast of Cibola". But also to the southeast should be found Marata, according to Fray Marcos, and indeed it is. The great ruin known today as Snaketown fits perfectly. Marcos also stated that Acus and other cities even larger than the seven lay beyond Cibola, and that points to the valley of the Verde River as the likely site. This valley is narrower than the others but very long, and ancient ruins have been found along nearly the entire length and in many of the side canyons as well.

A simple map of the region is given in Figure 3 where the locations shown for the Seven cities have been taken directly from Benham's map on the end papers. We see that the territory occupied by those people was immense, so as vast as was the picture painted by Fray Marcos of the Cities of Cibola and their neighbors, that picture easily fits within the frame at hand—a remarkable fact, and one not to be set lightly aside. How could it have happened, then, that his seemingly obvious association was not eagerly acclaimed from the beginning by everyone concerned with the matter?

The problem was fundamentally one of dating those old remains. Archaeologists today have very sophisticated means for determining the ages of cultures and artifacts, but early investigators were obliged to depend upon more indirect lines of reasoning. Often they were very indirect indeed, and that left room for a certain subconscious force to operate which could have colored their judgement somewhat. That is, we

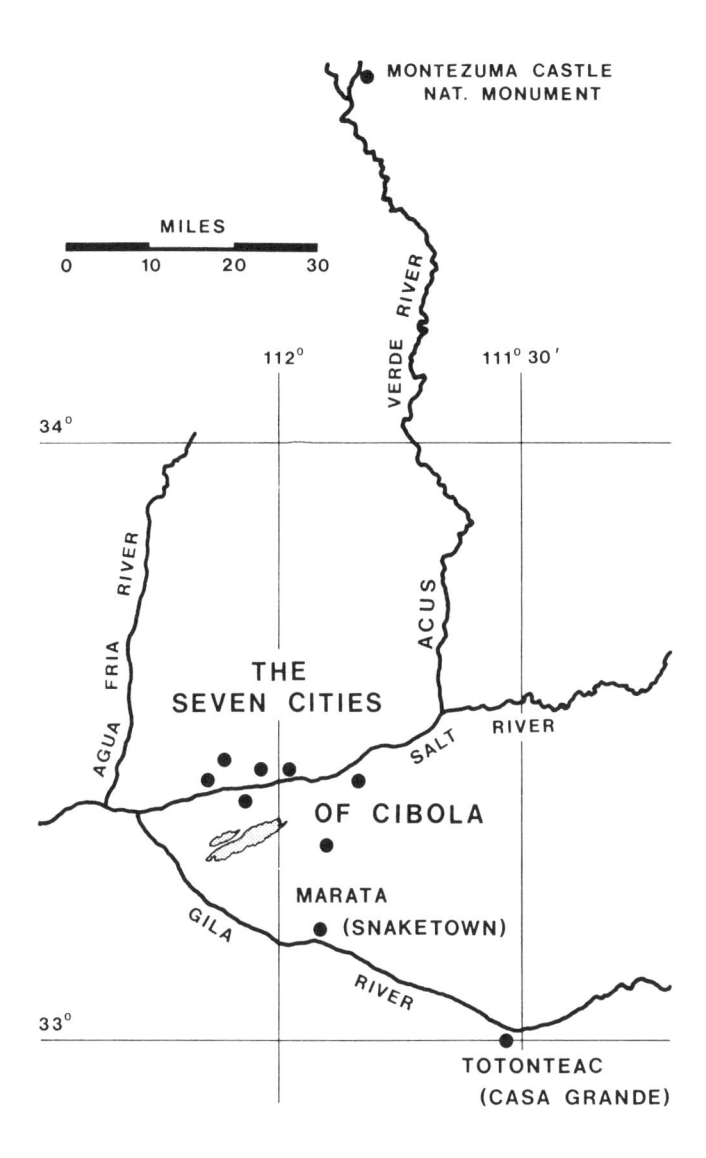

FIGURE 3: *Geographical distribution of ancient cities in the valleys of the Verde, Gila and Salt Rivers, in perfect agreement with the description given by Fray Marcos de Niza.*

understand that the goal of an archaeologist is study the roots of human culture, and to this end the very oldest remains would be of the most interest. And perhaps one can recognize a converse aspect to this yearning as well. In his enthusiasm, might there not be a tendency to estimate the greatest possible antiquity for the site at hand? Certainly no responsible investigator would knowingly falsify or misrepresent his data, but in the absence of definitive evidence those clues which seemed to indicate great age might in all honesty be given greater weight than others that point in the opposite direction.

We may be able to discern this force at work in the present instance by going back to the time of those first excavations and attending to the train of reasoning almost from Cushing's own lips, despite the fact that he was not privileged to speak through the official report of the Peabody Museum. For it happened that a reporter from the San francisco EXAMINER visited the site, and in a series of lengthy news stories he fairly well described the proceedings. The rationale by which Cushing estmated an age for the Los Muertos culture is itself interesting, but other facts are mentioned along the way that are valuable as well. So let us now read what the citizens of San Francisco read in their newspapers on the morning of January 22nd, 1888. The headline reads "THE SEVEN CITIES" and the reporter writes as follows [70]:

> " ... When Cushing first began his work he was under the impression that he was dealing with but one city. In need of a name, he called it Los Muertos. But as the settlers, attracted by his investigations [and] the fertility of the valley, began rapidly to clear their land, the discovery was made that with not one, but with many cities the investigator had to deal. By the time this fact was thoroughly ascertained, the essential character of the first city excavated, Los Muertos, was determined. It presented, indeed, a singular appearance. It consisted of an aggregation of large pueblos, of blocks, closely grouped together around a central temple or citadel building. Each pueblo was

Dead Men

capable of accommodating from 1000 to 4000 souls, according to its size."

Here follows a discussion of the clan distinctions among the modern Zunis and strong similarities to the same which were found in these ancient ruins. Then he describes the oven and funeral pyre, which is somewhat remote from the present topic, but after this he continues:

" When Mr. Cushing was living at Zuni, Professor Adolph Bandelier, now the historian of the Hemenway expedition, was pursuing his ethnological studies in that quarter. Mr. Cushing, from his initiation into the Zuni tribe and priesthood, was recognized as an almost absolute authority in all matters concerning this interesting and primitive people.
" Among other questions, Professor Bandelier asked him this one: "Why do the Zunis speak of the masters of the Six Great Houses. The Zuni town is one gigantic pueblo. What do they mean with their Six Great Houses? ...
" ... All that I can tell you", said Mr. Cushing, "is that there are Six Masters of the Great Houses. One is Priest of the North, another is Priest of the South, another of the West, another of the East, a fifth is Priest of the Under-World; the sixth, the Priest of the Over-World or skies. These six men, together with the Priestess, constitute the Supreme Council of the Zuni Tribe. ...
" The surprising fact, which it is the pleasure and the duty of the EXAMINER to communicate, is that the Great Houses of the Zuni Priests have been discovered. It is as strange and as interesting a fact as science affords that among the pueblos which constitue the cities of the Los Muertos system rise the six Great Houses of the ancient priesthod. Nay, there are seven houses corresponding to as many cities. The temple building, which is the central edifice of Los Muertos, Los Hornos, Los Pueblitos and the other cities, is the Great House of the ancient Priesthod. To the astonished members of the Hemenway expedition, this fact was at first regarded as startling coincidence, but a still stranger discovery increased the wonder. A hasty

investigation of the great ruin near what is now Mesa City, revealed the further fact that in these ruins alone of the entire seven, there were not one, but seven temple buildings! ..."

The reporter goes on to elaborate on this system of sevens and its supposed mystical significance, and then he comes to the question of the age of these ruins:

" ... How old are the curious skeletons, pictures of which have appeared from time to time in the EXAMINER? Curiously enough, we are actually able to form an approximate idea of their antiquity. Seven miles east of modern Zuni there is a curious group of seven cities which give us the strongest kind of proof of antiquity. A volcano eruption has filled the once fertile valley with lava. Against one of the cities the lava flow ran up on to the mesa and actually against the walls. The same flow, continuing down the valley, entered the mouth of the Zuni canyon. And just there is a curious fact. After the volcano had spent its energy and destructive force, after the molten rock had cooled, a living stream formed and rushed down the narrow opening of the Zuni canyon. The gentle action of the stream has, during the centuries of time, finally succeeded in cutting through the solid basalt at least four feet. None but geologists would realize how much is implied by this phenomenon. Taking the most conservative standard, that which Lyell applied to the Niagara Falls, not less than 6,000 years would be required to do this work. The pueblos, having been built before the lava flowed, must have been at least 6,000 years old and probably much older. But the type of their architecture is far more modern than that of Los Muertos, and these ruins occupy a position much further south in the course of migration. Therefore, we are furnished with many cycles of time with which to account for the rise of the cultures. Who shall say that civilized man in America is not ten thousand years old? The discoveries in the Salt River valley justify the statement. It is not easy to controvert the force of such geological evidence as is presented by the wearing down of the little stream in the Zuni canyon through the solid lava rock."

Dead Men

These passages have been quoted at length because they contain much firsthand testimony concerning the Hohokam ruins which is otherwise not available and because they give us some insight into Cushing's reasoning processes. Notice that the writer concurs in the number of cities. There were seven of them. We also learn that the ancient city which occupied the site of the present city of Mesa was the proud possessor of seven of those Great Houses and was therefore most probably the headquarters for the lord of the Seven Cities—that is, Ahacus, which Marcos described in Paragraph 13 of his narrative. This would be the easternmost of the seven as shown in Figure 3, or the one labeled "F" on Benham's map.

Let us recall that only one of those seven cities was actually named Cibola, and perhaps we can identify which one it was from the fact that it was the first city encountered by travelers who crossed the desert coming from the south. Judging from Figure 3, one would probably expect them to have arrived first at one of the cities on the Gila River instead. But they did not, so apparently their course lay somewhat to the west, and their final approach to the seven cities was from the west as well. Presumably their route was fixed by the availability of water along the way, or it may simply have been that visitors were less welcome at Marata and Totonteac. Whatever the reason, the city named Cibola was most probably the one south of the Salt River, near the western end of the group—the one that Benham labeled "E". In that case, Fray Marcos could have looked down upon Cibola from the South Mountains (shown shaded in Figure 3) in agreement with his statement in Paragraph 23. This would also locate Marata directly to the southeast of Cibola, in agreement with the information that Marcos gave in his Paragraph 13. Otherwise, one might have thought to identify Cibola as the city closest to his presumed inscription—the one Cushing later dubbed Los Muertos, just east of the South Mountains. But that would place Marata directly to the south.

Now as we learned from the reporter above Cushing was an initiate into the Zuni priesthood, so his opinion about the meaning of those sevens might well have been correct. Since the artifacts that he uncovered had so many familiar features there may indeed have been very close ties to the Zunis; after all, there had to be some reason why there were six masters of the Six Great Houses. But having followed his train of reasoning we can easily see why he did not think to identify that metropolis with the Cities of Cibola. For one thing, he had overestimated its age by perhaps 90 centuries.

Forty years passed, and Omar Turney, a former City Engineer, published a detailed study of the old canal system. Dr. Turney's lifelong specialty had been hydraulic engineering as applied to irrigation so his studies were especially thorough. In Part II of his report he made frequent reference to another very interesting fact that influenced early attempts to date the early culture. He pointed out that the heads of those old canals now stood high and dry as much as eighteen feet above the water level in the river! The unavoidable conclusion was that the bed of the river had been eroded down by those eighteen feet in the space of time since the canals had been used. Here is an example [79;p.20]:

> " ... In 1902 Canal Twelve remained just as it had been abandoned a thousand years or more, its head coming to the river bank and looking down eighteen feet to the sparkling surface below; a canal left high and dry on account of erosion. This canal is now paralleled by the modern St. John's which equals it in length."

A word or two of explanation may be in order before continuing with this line of the discussion. First of all, Turney had his own system of numbering those canals. His Canal No. 12 was the one Benham had earlier called Ancient Canal No. 1. Next, as already mentioned, the early settlers irrigated from the river and even made use of some of those same old canals.

Dead Men

How did they do it if the river had eroded down so far? The method, simply, was that the new canals left the river at a very slight angle and ran nearly parallel to the river for some distance, but at a smaller grade than the river itself. Because of that smaller grade the canal water gradually gained elevation over the level of water in the river until eventually it could be directed inland where it could be made to flow in one of those ancient conduits. However the old canal thus served would originally have taken its head far downstream from the head of its modern feeder.

But the ancient farmers did not have to resort to this strategem, for as Turney notes explicitly [p.16]:

" ... All the canals start out nearly at right angles to the river, and run directly to the lands to be served; they give no evidence of having been run up the river bottom to a higher grade."

This great drop in the level of the river bottom speaks for an enormous span of time since the old canals could have been utilized. In order to test this conclusion Turney consulted with a number of eminent authorities in this field whom he conducted to the river for an on-site inspection, and then he reported their impressions as follows—let us keep in mind that the river bed was dry at that time since retaining dams had been built upstream. Even so, the former water level could apparently still be discerned in those days [p.36]:

"   Standing in the boulder bed in the river, looking upward to the open channels of System Two above on the bluff, Dean Cummings, Professor of Anthropology of our State University, said, 'It seems as though two thousand years were too brief an estimate of the time needed to create this change.' and then thoughtfully added. 'It is not enough.' The Dean of our College of Mines and Engineering, Dr. Butler, a geologist, examined them and said, 'The estimate [2000 years] is reasonable, very reasonable.' The Professor of Astronomy, Dr. Douglass, the world authority on the record of tree rings, by using their testimony states that the abandonment might have occurred at the time of the later drop in rainfall between 500 A.D. and 600 A.D. or that it may have been before the beginning of the Christian Era. The Professor of Geography of Northwestern University, Dr. Haas, has said, 'This river aggrades nearly all the year and degrades only during the short time of high water, the net degradation is small; probably more that two thousand years have [been] required for such channel erosion' ... That French trained specialist on paleolithic man, Dr. Renaud, of the University of Denver, standing in the eroded river bottom and looking at the mouths of System Two, summed up all the conflicting lines of evidence and stated his opinion that these canals could not have been used for fifteen hundred years. After all came Dr. Marvin, then President of the University of Arizona, and said, 'All these estimates are far, far too recent, these canals came nearer being coeval with the power of the Pharaohs of Egypt.'..."

So this great erosion of the river bed is seeemingly even more convincing evidence for the great antiquity of that former culture. Perhaps one should not be surprised after all that these old ruins were not quickly identified as the Cities of Cibola, which were said to be alive and thriving in the year 1539.

As is well known, several objective and presumably absolute means for dating early cultures are available to archaeologists today. The radiocarbon method is probably the most widely familiar, and it is often supplemented by a highly developed technique for correlating the growth rings in trees and timbers. However these modern methods have only added to the confusion surrounding the Hohokam because they show without a doubt that those people lived until a very much more recent time than had been formerly imagined. In fact, their demise is now dated somewhere in the vicinity of A.D. 1400! Now the magnitude of the problem becomes all the more evident for we have just heard a whole panel of learned men

Dead Men

deduce from the eroded condition of the river bottom that probably two thousand years have passed since those canals could have been utilized.

But an even more recent date for the fall of that nation is required if they actually were the same cities that Fray Marcos described. Accordingly, we must now ask how firmly those objective methods determine the specific date A.D. 1400 given above. It turns out that tree ring dating has been only marginally useful in the Salt and Gila River valleys since few trees and worked logs survived to be tested. It is probably a small loss since dates determined by that method must pertain most directly to the building stages of a community; they would correlate with its fall only by inference. Radiocarbon analyses, on the other hand, should be useful in dating all phases of the culture—even the very end of it. However archaeologists find the radiocarbon results to be strangely puzzling in regard to the Hohokam. Let us hear a long-time expert tell of his own experience with the technique. Professor Emil Haury is here speaking of a group of 32 radiocarbon assays made on artifacts taken from Snaketown (that is, Marata according to our interpretation) during the years 1964 and 1965 [43;p.333]:

" ... It would be an understatement to record that the results were in agreement with each other. The opposite is the case. The task of sorting out those dates that appear usable from those which are obviously incorrect and justify one's selections is not simple. It is unthinkable, however, that dates for the Vahki Phase materials as far apart as 425±115 B.C. and A.D. 1020±120, or for the Sacaton Phase of A.D. 900±100 and 1820±110. can all be correct. These discrepancies force a choice. To do otherwise would land us in a chronological quagmire.

" The reason or reasons for these disparities may be many, ranging from the selection, collection, and recording of samples in the field, to contamination, and analytical errors in laboratory processing, and even to the assumptions on which isotope dating is based. It is not my intent to try to determine where the problem lies. However, I firmly believe that in making a qualitative judgement about the value of dates, an intimate knowledge of field problems and of the nature of the cultural complex under study is fundamental to the decision. ... My dependence on certain radiocarbon dates and rejection of others will not be pleasing to everyone, but these judgements must be made. If the complications arising from the establishment of a chronology of Snaketown ... have taught us anything, it is that we are far from having reached a finite level of expertness in the art of dating as applied to archaeology."

Evidently the objective scientific methods of dating are not so objective after all. Haury tabulated the results of the remaining 31 of those 32 radiocarbon measurements, omitting the Sacaton Phase sample which had indicated a date of A.D. 1820 ± 110 years. Now Sacaton is the name given to the most recent phase in the culture at Snaketown, and fortunately his list contained three more samples from that phase. After making the small correction implied by the tree ring calibration he gave dates for these three of A.D. 935, 965 and A.D. 1660. Prof. Haury estimated that the last date given may be accurate to within a hundred years as is customary for good samples stemming from the 17th century.

Of course, one would not expect all the samples to give exactly the same age because the various phases represent intervals of time, not specific dates, but the wide variation in these results does seem anomalous as Haury plainly states. Later on we shall find a plausible accounting for these discrepancies, but for the present let us merely note that these findings do support a substantially more recent date for the fall of those cities than is presently acknowledged. In the light of Haury's illuminating remarks above, then, it is easy to conclude that the date A.D. 1400 presently given for their demise is but an uneasy compromise between the objective evidence newly at hand and the older ideas of extreme antiquity.

Dead Men

Now one can understand how it happened that early scholars never thought to identify the Hohokam cities with Cibola even though they answered the description so well. Firstly, of course, Marcos himself had publicly identified the Zuni pueblos as the cities he had seen before so the true cities of Cibola were not even candidates for discovery. And then also the age of the Hohokam civilization was vastly overestimated from the beginning. Of course, Cushing's estimate based on an assumed rate of erosion of an unrelated lava flow in New Mexico was entirely unfounded, but other indications of great age had to be taken more seriously. Namely, by actual observation up to 18 feet had eroded away from the river bed since the days when those cities could have flourished, and in the normal course of events that great change would have required several thousands of years at the very least.

But what about earthquake? Perhaps the land rose by 18 feet during some such cataclysm; then the river would erode its bed to a new depth in making its way to the sea. As a matter of fact Cushing himself was of the opinion that earthquake had brought an end to the cities, but he thought that the people had only migrated to some other home because of one. According to his theory those Indians had a deep longing to live at the perfect center of the world, and since that ideal spot would presumably be free of earthquakes then even a minor temblor would be seen as proof that they had not yet arrived at that perfect place; they would move on in search of it. But where could they have gone? We recall that Cushing was an initiate into the Zuni priesthood, and that may have been the reason why he looked for a sublime interpretation of the signs when a more ordinary one would bave brought him closer to the truth. We need not agree with him on this point, but let us note his reason for thinking that an earthquake had occurred nevertheless. This detail also is to be found only in the newspaper, so let us rejoin that same reporter in that same edition of the San Francisco EXAMINER as he explains that supposed migration.

"   Mr. Cushing's explanation for this phenomenon [the migration] has not perhaps been thoroughly understood. There is an impression gone abroad that the City of Los Muertos was destroyed by earthquake and the inhabitants killed. Such a theory is essentially ludicrous and sensational. There is no doubt that the walls of Los Muertos were severely shattered and often thrown down by great earthquake. There is no doubt that many of the inhabitants were killed by falling walls. An illustration of such an accident is given in the EXAMINER today. But, serious as was the disturbance and unfortunate as were the consequences of the earth's upheaval, there was a far more important reason which impelled the migration of the people, a reason which sprang from the very bowels of their philosophy and religion. ..."

We need not pursue it. The point is that it was easy to see that the walls had been thrown down and that people had been crushed under them. A drawing made from a photograph of one of the victims accompanied the reporter's words. We had already deduced from those conflicting signs at the Casa Grande that the cities had been physically destroyed, and here are first-hand observations which confirm the fact. We had also deduced from the name Hohokam itself that the cities' residents had been killed on the spot; here again is corroboration. Surely one cannot imagine that the overturning of those walls occurred at any other time than the last days of that city. If the city had survived the episode then the rubble would have been cleared, the walls would have been repaired, and the victims would have been burned or buried according to the customary rites as a comfort to their families. But none of these thing was done. The walls lay where they fell.

It is interesting to note that although Cushing himself discounted the idea that an ultimate catastrophe had overtaken the nation ("Such a theory is essentially ludicrous and sensational"), others who were present and who examined the evidence for themselves did not take such a restrained view. Why else would the reporter have written: "There is an impres-

Dead Men

sion gone abroad that the City of Los Muertos was destroyed by earthquake and the inhabitants killed."? Modern archaeologists discount these signs altogether because Arizona is not subject to severe earthquakes. Gentle disturbances have been felt on rare occasions, but at least within recent times it is unlikely that serious damage to man or property has ever resulted from one.

Furthermore, if one considers the evidence available today then he must conclude that no such wide-scale shift has occurred, because a variation in the level of the land by eighteen feet would certainly have rendered the courses of those former canals invalid. Now it is true that the evidence was already far from complete by the time careful note was taken, so many of the routes shown on Benham's map had to be deduced from mere frgments of the old canals and knowledge of the present topography, but some of the ancient conduits were put to use by the early settlers almost as they were found, and they worked perfectly well—proof enough that no general shifting of the terrain had taken place during the intervening years. The old courses, then, remain valid routes even today —in most cases, that is, and one of the apparent exceptions makes an interesting problem in its own right. Let us consider that one now.

Omar Turney, we recall, was a specialist in irrigation systems, and he was greatly interested in that ancient civilization whose ruins were then so plainly evident on all sides. Consequently it was only natural that he should have thought to test those early farmers as hydraulic engineers. With this end in mind he systematically measured the form and grade of the old conduits in order to learn how well the ancients understood the principles involved. Most of their canals were very well designed indeed; modern engineers could do no better, but in one instance he encountered a remarkable puzzle—it seems that the old conduit had essentially no grade whatever, yet it ran for more than 12 miles! This was the northernmost canal of the old system in the western portion of the valley of the Salt River—the one that Benham had called Ancient Canal No. 5.

Turney compiled and published a map of his own which went through five editions, and upon that map he pointed out this strange feature explicitly. Plate 7 is an enlarged segment of his 1922 version, and Plate 8 shows the same region on his version of 1929 [79]. Notice that in the main he reproduced the course that Benham showed for this mystery canal, even to the branching features around that small knoll. The two versions of his map are almost the same, but they differ slightly with

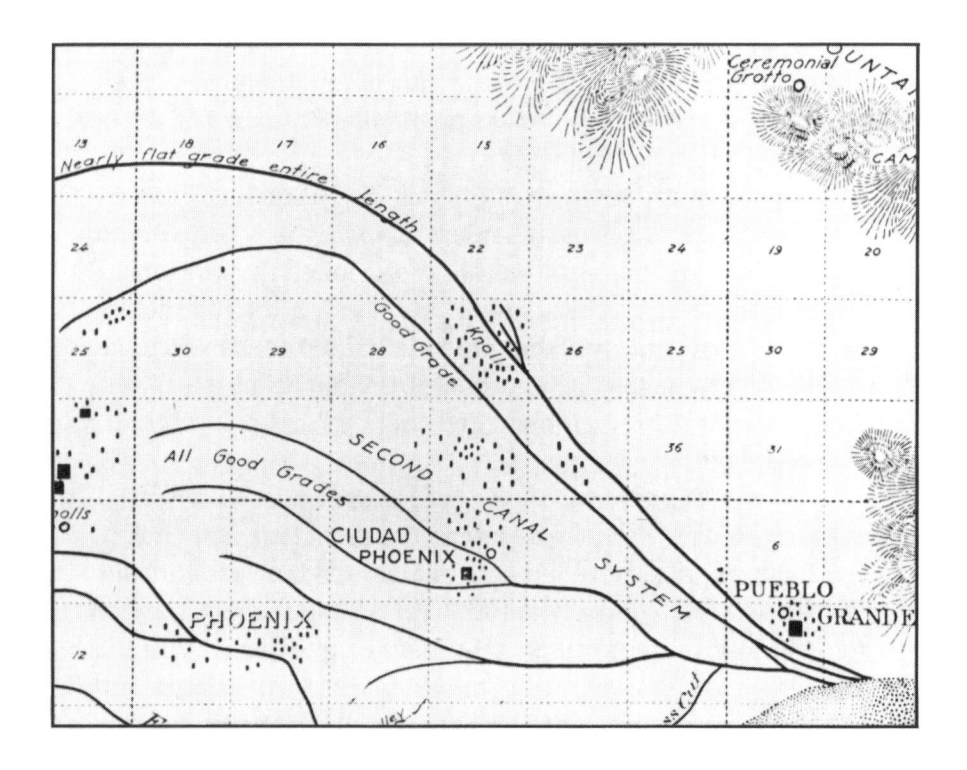

PLATE 7: *A segment of Omar Turney's 1922 map of the prehistoric canal system in the valley of the Salt River, apparently unpublished. This photograph was made from a blueprint of the original drawing, identified only by the copyright notice.*

Dead Men

respect to some details near the Pueblo Grande. By eliminating one of the branches there he gained a canal head at the river, and in 1929 he called that site the "Park of Four Waters" in honor of this new addition.

Plate 9 is an enlargement of the corresponding portion of Benham's map. Let us note in passing how these two authors showed the courses of modern canals; Turney indicated them with dashed tracings while Benham used parallel solid lines. Although the names of the modern conduits do not appear in these small segments, one can easily recognize the modern Grand Canal as the one passing close to the Pueblo Grande

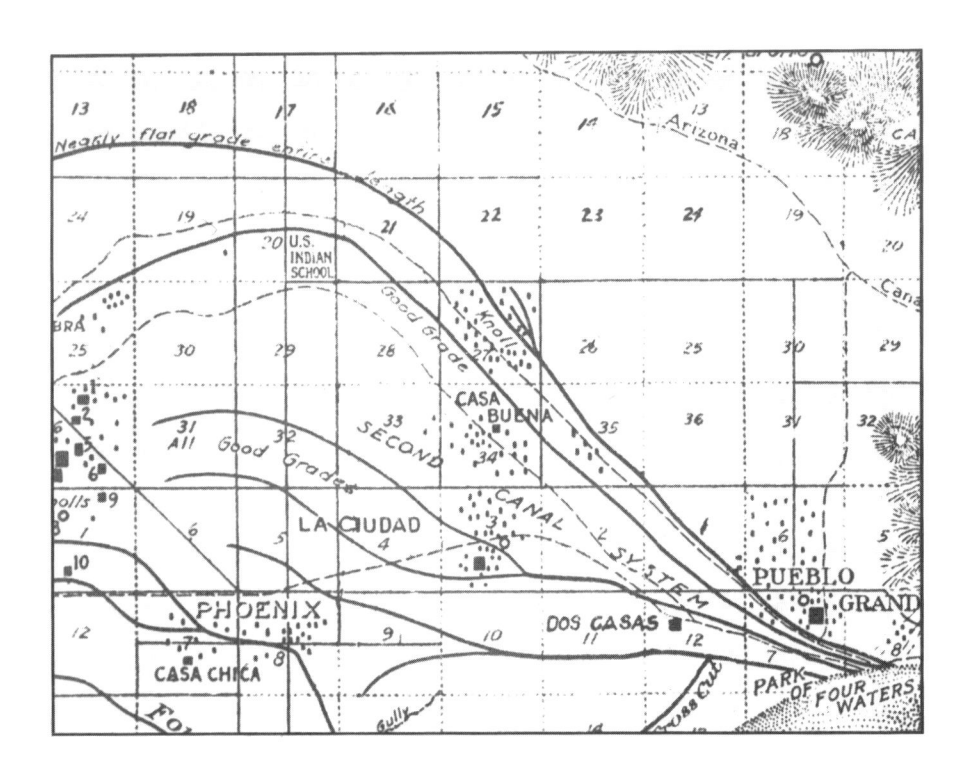

PLATE 8: *The same region shown in Plate 7 as rendered on Turney's map of 1929 showing notable changes near the Pueblo Grande.*

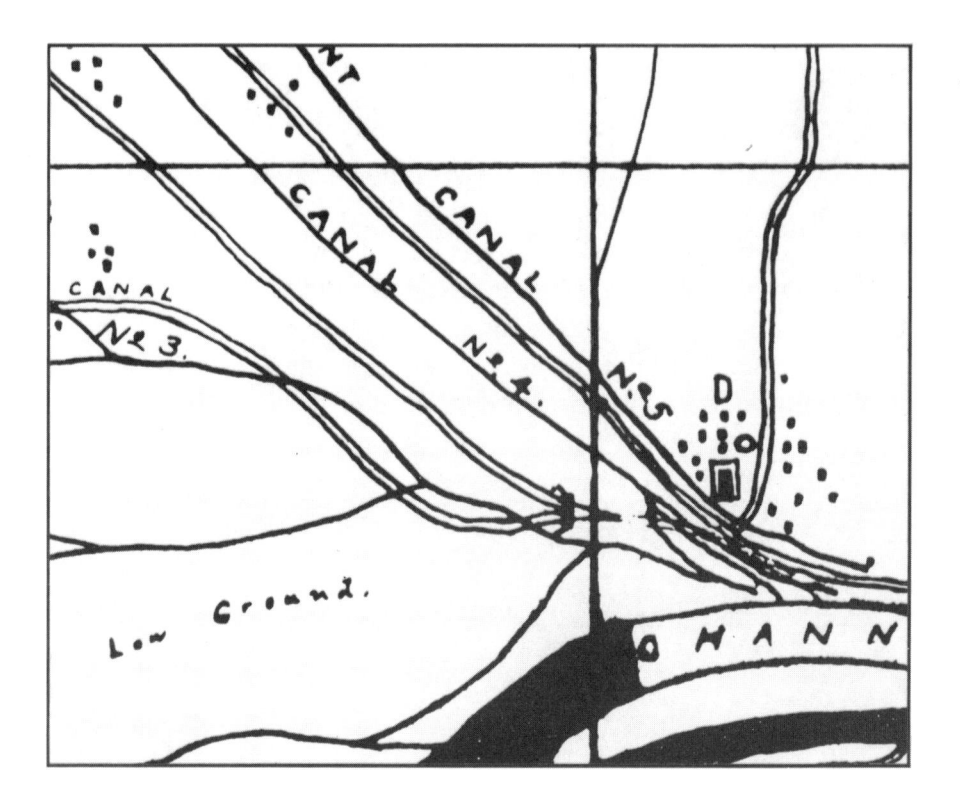

PLATE 9: *An enlarged segment of Benham's map of 1903 showing generally the same region as in Plates 7 and 8.*

(compare Plate 3).

Now as one studies Plate 9 he must notice a very strange thing, namely, that Benham showed neither three nor four, but only two canal heads at the river in this vicinity. There appears to have been a minor question about Canal No. 3 so he left it incomplete, just short of joining with the others. But this Mystery Canal No. 5 comes to an abrupt end well away from the river with no prospect of a head whatever. This seems extremely odd for although the ancient system was far from complete in his day, Benham did not hesitate to fill in missing portions in other areas. Why should he have hesitated here?

Dead Men

Apparently he was aware of some fundamental problem with the evidence—one that he struggled with but could not resolve. Let us try to reconstruct his thinking and attempt to deduce what that problem might have been.

Irrigation canals must be laid out in harmony with the topography so let us examine the constraints at hand. First of all, the valley of the Salt River is a gently sloping bottom land most of the way downstream from its junction with the Verde to its end where the Salt empties into the Gila (Figure 3 shows the lay of these rivers). One notable interruption in this regular terrain is found in a narrow span of rolling, rocky ground on the eastern reaches of the city of Phoenix; Plate 10 is a view towards the south looking down upon this area. Notice that it is dotted here and there with a number of small buttes of an uncertain character. Since the land is useless for agriculture it has never felt the plough, and much of it remains in its native condition even today. A large portion has been set aside for recreational purposes, having been named Papago Park after a nearby tribe of Indians. This elevated region extends all the way to the River just a few hundred yards upstream from the Pueblo Grande.

Benham, then, presumably had in evidence residues of that northern canal, and he set out to reconstruct its course back to the river so that he could enter it on his map. His task should have been an easy one. Given the grade required by a functional irrigation canal, he had only to trace backwards along such a grade from the last clear residue until he arrived at the river.

But that higher rocky ground must have presented an insuperable problem. In principle he might lay out a plausible route to a head sufficiently far up the river, but the virgin surface in that region testified loudly against it. That ground had not been disturbed, either by the modern settlers or by the Indians before. To avoid that obvious conflict with the evidence, then, he picked out the route shown, but in doing so he

PLATE 10: *A view toward the south looking down upon the span of rocky ground at the eastern boundary of Phoenix.*

arrived at the river too far to the west, that is, too far downstream and at much to low of an elevation.

In later years Turney analyzed the problem somewhat differently. He was willing to assume a head for the canal in the vicinity of the Pueblo Grande and simply acknowledge that the conduit did not have adequate grade. He thought that the Indians were so desparately in need of more land for agriculture that they were willing to pay the very high price for water that this fault implied. But would this option even have been available to those formers of old? It is not likely because there can be no flat irrigation canal. These vessels are required to transport water, not merely contain it, and they need an appreciable slope in order to be useful at all.

One ever-present problem that arises with a canal of this kind is that silt tends to settle out of the water if it is permitted to flow too quietly. The modern canals are built with low waterfalls at intervals along the line to generate turbulence in the stream so its burden of silt does not settle out. The ancient Indians fixed rocks into the walls of their canals to generate turbulence for this same purpose. But this artifice requires an appreciable rate of flow to be effective. Moreover, those canals had no lining so water continually seeped out through the bottom along the way. Therefore a substantial rate of flow would be required near the head merely to supply this leak in a canal that was twelve miles long.

In plain words, then, the two maps agree on this remarkable testimony: In the early days, when the signs were fresh, a segment of an ancient canal remained in evidence which was entirely cut off from the river. That is, there is no feasible route consistent with the present contour of the land along which that downstream segment could have received its water. The problem is so clearly defined that its solution is all but forced. We can now be only one step away from understanding, but before taking that last step there is one additional item that must be considered.

Although nothing whatever remains of that mysterious canal today, some tangible evidence relating to the problem has survived. As it happens, of all the many miles of ancient canals which once coursed over the Salt River Valley only three short segments still exist, and one of these is at the so called Park of Four Waters. The relics can be seen in Plate 11, this being a portion of an aerial photograph made by the U.S. Geological Survey in 1967. Let us note some of its main points.

The curve along the very bottom is the northern edge of the river bed. Those black segments on the right, at the edge of

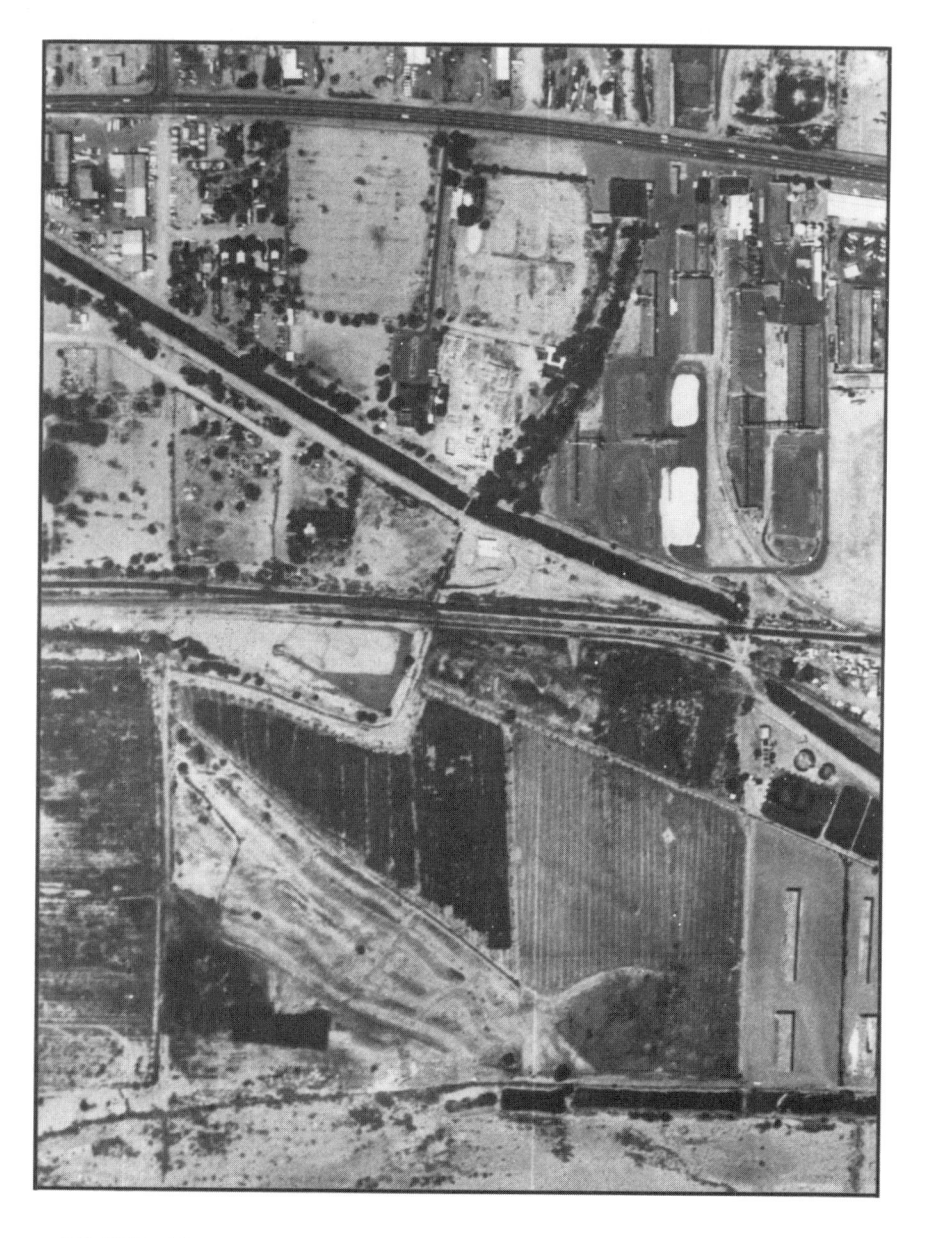

PLATE 11: *A segment of an aerial photograph made by the U.S. Geological Survey in 1967 that shows the Park of Four Waters region (Photo No. 3-14-67; 1-148; GS-VBUL).*

the river, were sewerage treatment ponds in former times; the ancient canal residues are the undulating features that border the river to the left of those ponds. Notice especially the pronounced "spur" that extends out to the right from this undulating region. Further to the north, near the center, one can discern the railroad line running east and west, while the Grand Canal passes under the tracks and works its way to the northwest. The old crosscut canal curves into it just above center, and one can make out the Pueblo Grande residue there at the junction, just above the very small bridge across the canal. Finally, then, Washington street passes across the top of the picture.

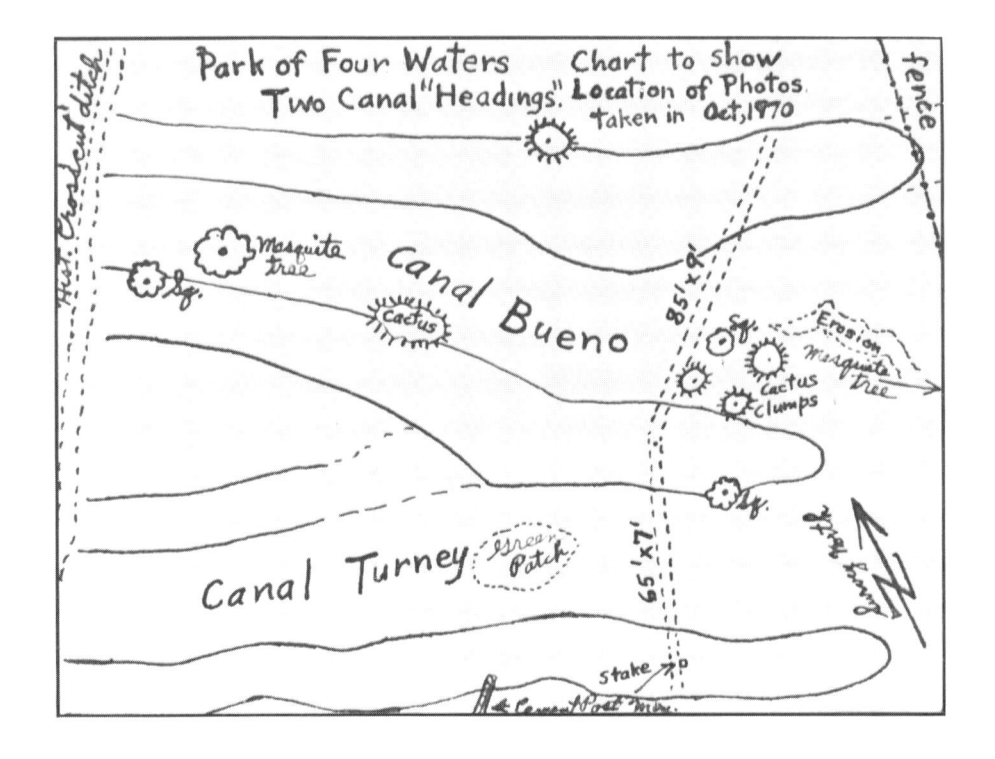

PLATE 12: *A sketch by the late Frank Midvale of the ancient canal residues at the Park of Four Waters.*

The old residues have suffered much during the intervening centuries, both from man and from the elements, so they are difficult to interpret even on the spot. Plate 12 explains them nicely. This sketch was drawn by Frank Midvale, a long-time student of the old canal system, shortly before his death. Midvale's personal papers and notebooks are now in custody of the Department of Anthropology at the Arizona State University in Tempe, and this drawing was photographed from amongst those papers. Notice that Midvale did not indicate the spur-like extension visible in Plate 11. This also has fallen victim to the times. It was recently destroyed in order to make way for an overflow sluice from the canal to the river bed.

At least today there are only two canal heads in that vicinity, and it is easy to conclude that most probably there were never any more. Presumably those two canals were joined at their heads so they could be fed together during seasons when the water was low. At such times the Indians would have been obliged to construct diversion dams across the river to raise the water level, and both of these heads could have been fed from behind the same dam. That spur-like appendage testifies that these were the actual heads of the canals. It would have diverted water into the canals, and it would help prevent flooding in the region to the north. One can now appreciate Benham's meticulous attention to detail all the more for it appears that he even indicated this spur on his map.

So it becomes ever more certain that the course shown on these maps for the upstream portion of Ancient Canal No. 5 cannot be correct. Indeed, the indicated course can only be understood as the veiled statement of an unresolved problem. It is also certain that no solution to this riddle is remotely feasible given the present lay of the land. Yielding to the evidence, then, only one alternative remains: Ancient Canal No 5 must have had its head well upstream from the Pueblo Grande, and it passed normally through the Papago Park region. *The high ground could not then have existed!*

Chapter 5:

SIGNS OF CATASTROPHE

FACED WITH THE mind-wrenching implications of that old irrigation canal we might pause here briefly to review the problem as it has developed to this point. Recall that when Cushing excavated the ruins at Los Muertos he found clear evidence that the ancient city had been destroyed by violence —even if he refused to admit it. Huge structures had been thrown to the ground, and people had been buried in the rubble. Cushing himself, along with succeeding generations of archaeologists, was blind to this obvious fact because of the fog that settled over those great cities when Coronado and Fray Marcos falsely identified the Zuni villages as Cibola. And being thus blinded they all lacked the courage to address the plain evidence forthrightly. Turney, for example, skirted the issue by inventing the "Park of Four Waters" out of whole cloth and supposing that the Indians were able to irrigate that vast area west of the high ground from a canal that had no grade. Likewise modern scholars imagine that the old culture flourished until the fourteenth or fifteenth century even though the river bed was eroded so deeply that their irrigation system was useless—and had been, according to sober testimony, since perhaps the time of the Pharaohs in Egypt.

But having cleared that fog away we now understand

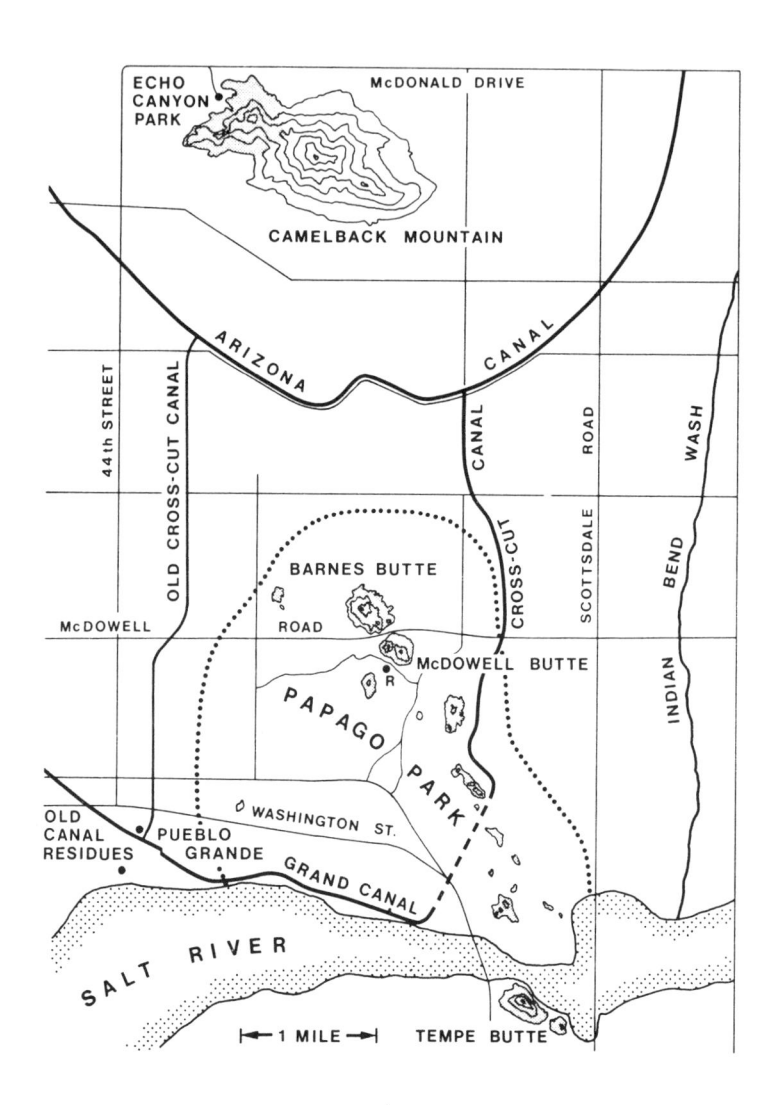

FIGURE 4: *Sketch map of Papago Park and relevant features in the surrounding neighborhood.*

Signs of Catastrophe

that those cities survived as thriving metropolis until only a short time before Father Kino arrived upon the scene in 1692. Certainly no familiar mechanism of nature could have wrought that great destruction so quickly, so let us gather the courage to admit that some heretofore unknown kind of scourge attacked the region and reduced it to ruin—and suddenly, leaving no survivors to tell the tale. In that case, we must keep our eyes open for the unfamiliar as we now enter onto that anomalous high ground to perform our own on-site inspection.

Figure 4 is a simplified map of the Papago Park region showing the features that have been discussed so far. As already noted, the river is now dry most of the time since retaining dams have been built upstream. Water is metered out as needed into trunk canals which follow along the northern and southern edges of the valley. We see in Figure 4 that the Arizona Canal serves the area north of the river. Water is delivered to the Grand Canal via a cross-cut canal passing down the eastern flank of Papago Park and through a tube under the high ground. This is how we happen to see water in the Grand Canal today even though there is none in the river. Benham's map shows where this canal originally had its head. It had to begin far upstream in order to supply water to approximately the same land that the Indians once irrigated from canals starting out just south of the Pueblo Grande. As already noted this strategem was required because of the intervening loss of those eighteen feet from the river bed.

Referring again to Figure 4, the dotted curve outlines the region where the terrain is most notably rocky and uneven. There is no distinct boundary, but the land does become more regular in the area beyond, and the elevation gradually decreases until it joins smoothly onto the normal contour of the valley. On the west side there is nothing to show where the deposition of new dirt might have ended; but on the eastern flank the slope of the elevated region opposes the general trend of the valley, and the two grades meet at a shallow arroyo which floods after heavy rains. This has lately been named Indian Bend Wash, and it is so indicated on the map.

We might note first that Benham shows on his map (reproduced here on the endpapers) two small settlements on this high ground—one represented by three small dots and the other by six. If these actually had been ancient settlements then the reconstruction offered here would be untenable. Whatever traces may have suggested the easternmost of the two are long since gone; that area is now completely urbanized so there is no hope of examining them for ourselves. However the other site lies on barren rocky ground and remains today largely unchanged by the encroaching civilization.

But this one can hardly be called a settlement. In fact, it is the residue of one small house as we see in Plate 13. The

PLATE 13: *Residues of a small dwelling on the high ground of Papago Park just north of the Salt River.*

Signs of Catastrophe

photograph gives only a vague indication of its size, but it was quite small, possibly the work of one man, and it appears never to have been finished. Clearly this isolation does not square with the Indians' customary style of life; they were a communal people. Although nothing of the walls remains in place a few of the foundation stones are still to be seen, and they betray a crude, light construction that further suggests the work of a single individual—one intent on satisfying his own limited needs. It has not the substantial character of the Indians' work, which aimed at permanency. All signs suggest, therefore, that this was the cabin of a lone settler in a frontier environment and

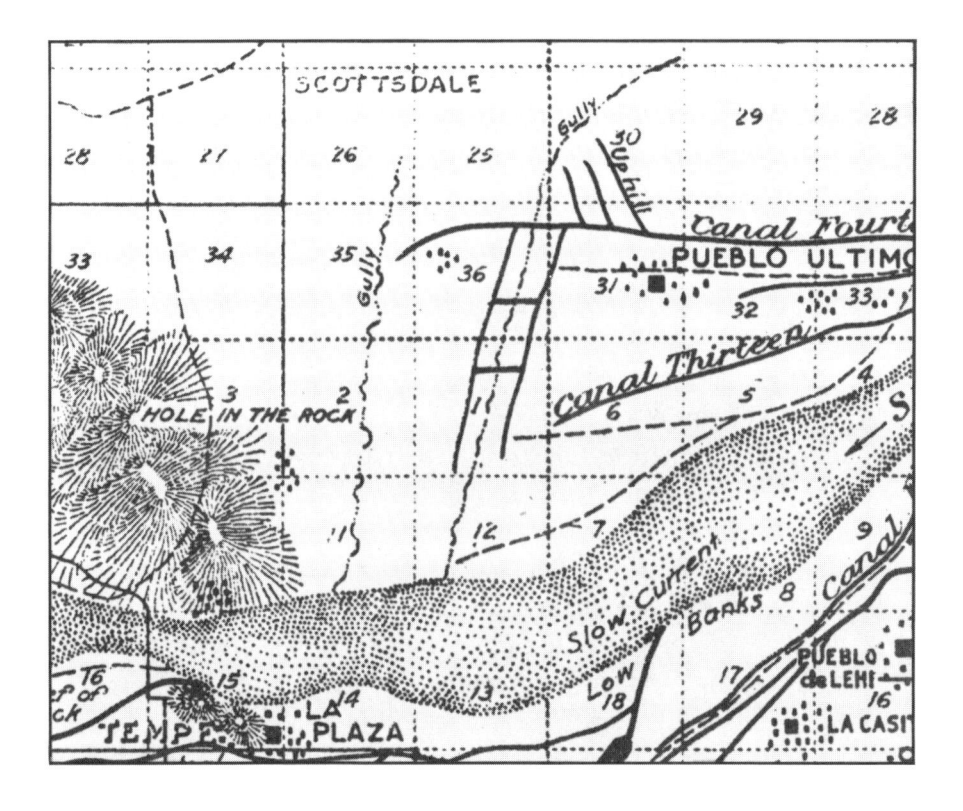

PLATE 14: *A segment of Omar Turney's map of 1929 showing the region to the east of Papago Park.*

had nothing to do with the ancient culture. However, it is interesting to note that if this high ground had existed in those olden times one might indeed expect the Indians to have built upon it since it was otherwise worthless. Building upon this ground would have freed tillable land for agriculture—but there was no such building.

Plate 14 is a portion of Turney's map of 1929 that includes this same area. Note that that cabin in now represented by eleven dots while the other "settlement" has grown to eight. It is quite evident, therefore, that these features were added to the map as mere embellishments and cannot be taken seriously for our purposes. Let us understand that those early observers had no suspicion that the terrain was abnormal so they did not think to make the fine distinctions that would be helpful for us today. Researchers in the future may want to sift through more of the sparce details that remain from those olden times with our present point of view in mind, but an exhaustive survey of this kind cannot be our goal here. Our picture of that period is already defined as closely as needed for present purposes by the facts already in evidence, so let us simply follow their lead to the end.

Continuing with Plate 14 while we are here let us note that one of those ancient canals (Canal Fourteen as Turney numbered them) after having followed a westerly course from the river came to an abrupt halt at the Indian Bend Wash. Of course, if the wash had existed in those former times then it could have gone no further, but why should the Indians have constructed a canal leading up to a wash that could not be farmed? This wash is a raging torrent today after even a moderate rain, and only grass and fair sized trees can grow there. Crop plants on a tilled surface could not. So the fact that this canal went so far west gives a broad hint that arable land also continued further to the west in those times.

And one can hardly fail to notice another curious feature upstream on this same canal; namely, three branches leave

Signs of Catastrophe

toward the north running uphill! The map points this out explicitly for only one, but presumably all three suffered from the same defect. Because of this fault, Turney was somewhat restrained in his praise of those old engineers when discussing these remarkable relics. He put it this way [79;p.18]:

" Clearly they were unable to determine where water would flow except by digging a channel and from such inability they could know little about the land to be reclaimed until completing the work. Let cease the boasting about ancient engineering skill; in few points only was it developed; a maximum velocity with the least earth removed was obtained by making the wetted perimeter bear a minimum ratio to the cross-sectional area. Long practice may produce results equal to technical skill."

Since we see that ancient world only dimly it is all too easy to imagine that the ancient people themselves saw their world only dimly and were not very bright. But if they were able to discover that the wetted perimeter of canals should be in minimum ratio to the cross sectional area in order to achieve the best flow for the least dirt removed then they should also have discovered how to lay out a course for their canals by digging small test ditches in advance of the main construction. In fact they probably did better than that. More likely they would have built small, clay lined ditches to define contour lines at regular intervals so they could lay out their routes—not merely downhill, but with a precise grade in view.

Perhaps one can account for this apparent blunder by noting that if a canal were buried by dirt sifting down from above all traces of the canal would be lost when the covering became sufficiently thick. But on the edge of the fall, where the added material was not so thick, the bottom of the canal, its levees and the surrounding terrain would all be covered to about the same depth. In that case the contour of the canal would still be evident at the surface; only its grade would be modified as the thickness of that covering would change. In fact, with respect to this new surface the visible channel would run uphill for a distance before it gradually disappeared—which is just what we see here. Although no source has yet been offered for all of that dirt, this does give a plausible accounting for the apparent folly of those old engineers, and it does so in the same context as our central theme.

One might note that Mystery Canal No. 5 would have presented a similar apparition to the early settlers. They would have been fascinated and perplexed to see two parallel ridges appear out of nowhere and proceed down the hill to gradually form themselves into a perfectly normal canal. But rather than waste time trying to resolve the unresolvable they set about leveling the land. We can only regret that they did not make a permanent record of that marvelous sight, but what would they have done with it? They were not archaeologists in the business of writing papers. They were farmers bent on preparing the land for agriculture—which they did, and all traces of that strange spectacle were destroyed.

Let us now return to the central topic and direct our attention to the elevated ground within the park region itself. Here we find a rolling, undulating topography that is interrupted here and there by the strange buttes seen earlier in Plate 10. The rock making up these hills would be classed as a breccia near the surface and a conglomerate somewhat below. That is, the surface mass is not a single unified rock, but many small ones that fit neatly together to form the whole. The size range of the component grains varies from one hill to another. In McDowell Butte the largest are granite blocks that must weigh several hundreds of pounds, while in some of the others they are mere pebbles. In all cases, however, the size grades down to a very fine silt that passes easily through a 300-mesh screen. Strangely enough, all of these elementary granules are individually coated with a thin layer of red pigment which is presumably some compound of iron. This ever-present pig-

Signs of Catastrophe

ment accounts for the dark, somber aspect of the buttes which is so plainly evident. When this pigment is removed the particles are found to be mainly light, quartzose grains, angular as if freshly crushed, and to all appearances commonplace.

Plate 15 brings us to our main region of interest with a view looking toward the northwest at the large hill just south of McDowell Road. Not previously named, this one has been dubbed McDowell Butte in Figure 4 because we shall need to refer to it often in the future. A portion of Barnes Butte is visible in the background, and Plate 16 is a view looking north at Barnes Butte alone. Upon noting the layered structure of

PLATE 15: *A view of McDowell Butte looking toward the northwest. Barnes Butte is visible at the left.*

this hill, one's mind tends to spring back far into the distant past and imagine detrital material collecting in vast seas which later drained away. The soft contours suggest prolonged weathering and erosion while the level of the land gradually changed, tilting the bedding planes upward. The shallow, cave-like holes that dominate the surface suggest a wind sculpturing process —erosion by the driven sand, again continuing through untold ages of time.

Such might be offered, in all fairness, as a "standard" interpretation of the features, but it could hardly be more seriously in error. If we observe the details closely we must find

PLATE 16: *Barnes Butte looking north. From this aspect it displays clear bedding features tilted up to the right.*

Signs of Catastrophe

that the rounded aspect of these hills had an altogether different origin; in fact, we shall discover no evidence of erosion whatever. As will be evident, these hills remain today exactly as they were the day they were formed!

Plate 17 is a view looking north into the small cleft in McDowell Butte from about the position of the picnic ramada indicated by the dot at the letter R in Figure 4. Again one sees the cave-marked, rounded surface, but notice especially the dome-like projection just to the right of center. There can be no doubt that it is rounded because on some past occasion a portion of the rock became liquified and flowed downward.

PLATE 17: *A small canyon near the western end of McDowell Butte, looking north.*

PLATE 18: *A closer view of the dome-like structure visible in Plate 17 showing superficial flow patterns in the rock.*

Plate 18 shows this same dome from a somewhat closer vantage point. One can also easily discern that a viscous mass oozed over the top of the cave at the upper right and congealed as it was about to fall. The bridge across the left hand end of this cave must have had its origin in a similar flow from above. Evidently some of the liquid continued downward, and the resulting rivulets congealed shortly afterward, even while they were running down the wall. Careful inspection removes all doubt; these really are the patterns of a flowing liquid.

Plate 19 is a view of the left side of the small canyon seen earlier in Plate 17; it seems that a fluid surface layer has been

Signs of Catastrophe

PLATE 19: *A closer look at the caves on the left side of the small canyon shown in Plate 17.*

added as a kind of "frosting" over rock already in place. Several places can be seen where gaps are spanned by relatively thin structures which must have been liquid at one time and which appear to be the congealed residues of that frosting which draped over the hill. The covering formed a kind of curtain over one cave near the center of the picture, but evidently it lacked the strength to support itself so it broke apart over the larger holes leaving residues that can still be seen clinging to the tops of the caves.

Plate 20 shows that curtained cave from the other side of the cleft. The presence of new material added over the old is

PLATE 20: *The "curtained cave" visible in Plates 17 and 19 viewed from the other side of the small canyon.*

very clearly revealed in this view. Also worth mentioning is the fact that although the surface at large displays clear evidence of fluid behavior the individual components show no sign whatever of having been melted. The sharp angularity of the larger rocks at the surface is plain even at a glance in this photograph, but this property is not limited to the larger rocks. All sizes, even the smallest grains, are likewise sharp and angular. Clearly, then, this was not a melting of the familiar kind, but a phenomenon of a different order altogether. In due course we shall infer a plausible basis for it, but for the present let us be content merely to explore some of its consequences.

Signs of Catastrophe

Perhaps one can point to this strange melting effect as the origin of the great strength of this mass of rock. As already mentioned, the grains at the surface are neither fused nor cemented together yet they form an exceeding hard and impervious whole because of their close fit. Nevertheless, with shearing forces, small pieces of the mass can be reduced to their elemental grains with hardly more than finger pressure alone. Presumably the grains fit so perfectly because at one time they flowed as if under pressure to fill the available space compactly, but they failed to actually fuse together because of the ever-present coating of red pigment on each one of them.

PLATE 21: *The northwestern region of McDowell Butte as seen from the slope of Barnes Butte.*

Let it also be noted that the sharp angularity of these superficial rocks speaks volumes for the youth of the structure at large, for no erosion whatever can be discerned in them. Even the delicate flow patterns, composed not of solid rock but of a frangible breccia, have survived almost in their pristine condition. And more significant still, the red coloration of the hills persists even though it derives not from color inherent in the rock, but from a mere coating on the individual grains. This point will be examined again later in another context.

Still other sites are readily accessible which display interesting flow patterns and also speak clearly for the youth of this added material so let us work our way around the hill and examine some them. In Plate 21 the camera was stationed on the southern slope of Barnes Butte and points in a southeasterly direction. Notice that rock decidedly in a fluid state has eased over the top of the right hand end of the cave on the lower right, and here, too, the fluid appears to have been part of a new layer that continued up the hill to form a partial curtain over one of the upper caves. Presumably it was a continuation of the same sheet that we observed on the other side of the hill. Further to the left the fluid appears to have been formed less like a sheet and more like two heavy "cords" which appear as bulges above the lower cave, and there is an obvious dribbling of the liquid down the back wall as well. Plate 22 gives a closer view of this dribble.

Certainly one can discard out of hand any thought that this feature might be an accretion structure resulting from the long-continued precipitation of minerals out of water trickling down the hill. Note that the cords are elevated *above* the general contour of the surface so they are not feasible water courses; water might collect and flow between them but hardly upon them. Furthermore, the texture is not that of a mineral deposition; indeed, it is not unlike that prevailing over the surface of the hill at large. Neither can one suppose that this nicely defined feature could have been sculpted by the degrad-

Signs of Catastrophe

PLATE 22: *A well-defined dribble of melt-rock running down the back of the cave near the lower center of Plate 22.*

ing processes of erosion. This is a small structure in itself, yet even its finest details remain perfectly preserved, giving added assurance, if any more were needed, that these hills remain today in substantially their original form. Nothing of significance has eroded away.

At least as far as can be observed locally, rocks of the type in question are unique to Papago Park, with only two exceptions. One of these, know as Twin Buttes, stands alone some two miles southwest of Tempe Butte, but since it offers nothing new it need not be considered further here. However

PLATE 23: *Pools of alien reddish rock solidified on the northern "neck" region of Camelback Mountain.*

the other locale does deserve some attention. It is the "head" region of Camelback Mountain, a prominent landmark at the northeastern limit of the city of Phoenix. The similarity of rock forms here to those occurring in Papago Park has been indicated by the shading of this area in Figure 4. For the present it will be sufficient to note in Plate 23 an area on the northern slope of the "neck" region which also shows an extent of newly added rock that was obviously once in a liquid state. This new rock might be described as a bright red sand-and-gravel-stone so it is of a substantially different kind than has been met previously. Plate 24 is a telephoto view of the region to the

Signs of Catastrophe

PLATE 24: *Telephoto view of an area visible in Plate 23 showing rivulets of rock running over the edge of the cleft.*

right of the steep cleft near the center where rivulets of this new rock are to be seen running over the side. The stark contrast between the fluid contours of this new red rock and the jagged aspect of the native grey rocks in the area is plainly evident. Localized occurrances of this alien red rock may also be found at the base of neck region on the southern side of the mountain. Here clear signs of melting in the usual sense are plainly evident, although they are not so easy to photograph because of the intrusive housing.

Now the dirt within the park is also very interesting in this regard since it, too, is of a reddish hue—and for the same reason as before; the individual particles are universally coated with that same red pigment. The color is less conspicuous upon the surface where the ground has not been disturbed, so some weathering is apparent, but even here the coloring in unmistakable. Only in arroyos where running water and mutual abrasion have accelerated the weathering process are clear, unpigmented grains to be found.

If running water and the resulting mutual abrasion of the grains removes the pigment, as one would certainly expect, then it must be obvious that *running water could have played no part whatever in depositing this material.* Both because this red coating is still intact and because there is no evidence of erosion from the hills this ground material must likewise be in its original pristine state.

Complicating the problem even further is the fact that this dirt is a distinctive mixture of silt and gravel in which sand-sized grains and the very small clay-sized particles are both notably scarce. This gives the grain-size distribution a bimodal character that can be observed widely throughout the park area. Normally one would expect the product of either a grinding process or a sorting process to be characterized by grains in a predominant size range, with a decreasing admixture of grains both finer and coarser. But here we find what can only be interpreted as the products of *two different* processes that were somehow intimately mixed, and the individual grains having been coated with pigment they were disposed upon the ground *without the aid of water.* Figure 5 is a histogram showing a typical grain-size analysis of dirt from the park area.

Without a doubt these residues testify to a phenomenon beyond the reach of known physics and chemistry. From the bizarre melting of the very rocks one might deduce that it was a violently energetic event—and also widespread, extending over more than a thousand square miles. We have as yet no clue as to the cause of the catastrophe or the source of the alien material covering that former landscape, but the unexpected

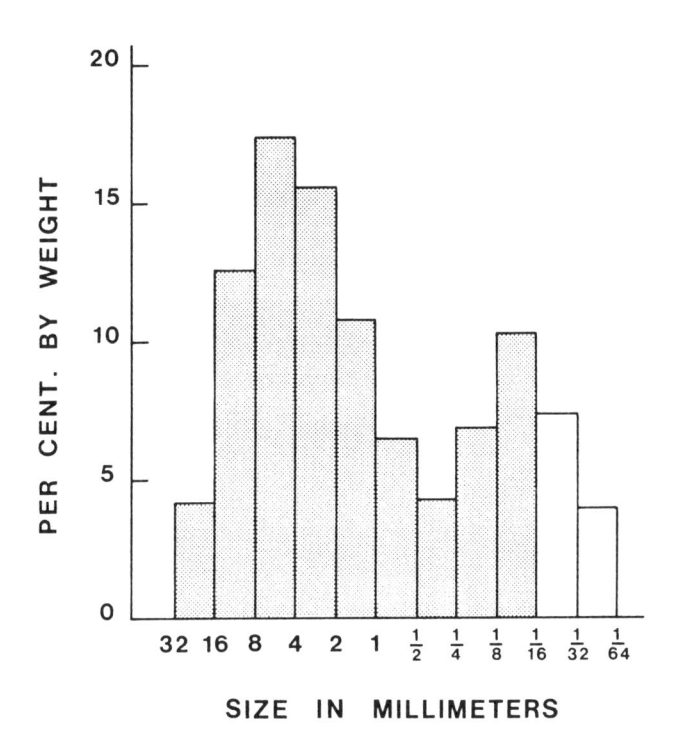

FIGURE 5: *Grain Size distribution in a soil sample from the north slope of Tempe Butte. The finer components, below 2 mm, were separated with the standard soil analysis sieves manufactured by Humbolt Mfg. Co. of Norridge, Ill. Below ½ mm the screens are not graded in fractions of a millimeter, although they closely approximate the sizes given in the figure. They are graded by mesh number—that is, wires per inch, the finest ones being numbered 60, 140 and 300. Two additional grades are shown unshaded. This is the portion of the sample which passed through the 300-mesh screen, and the allotment of this 11.4% into the next two grades is assumed. Obviously, the bimodal character of the distribution is very well defined however this allotment may be made.*

bimodal grain-size distribution observed in the dirt does suggest a pregnant next phase for study. Namely, it so happens that various residues from the ice ages also show a bimodal distribution in particle size, and this odd feature is no more plausible in glacial debris than in our present material. Is it possible, then, that those residues have been somehow misconstrued?

In order to investigate this question let us detour temporarily to the northern climes and examine some of those supposed glacial residues for ourselves. Then later, upon returning briefly to Papago Park, we shall be able to draw surprising conclusions from some of these rocky forms, and we shall also be able to account for those eighteen feet missing from the bed of the Salt River.

Chapter 6:

Ice-age Residues?

THE GREAT ICE AGES have been a familiar part of earth's history for more than a hundred years. Thousands of learned scientists from all over the world have examined their residues minutely from every possible point of view. They have plotted the several advances and retreats of the ice sheets; they have described various mechanisms at work, probable effects upon the ocean currents, the climate, the prevailing flora and fauna and any number of other details. Modern advances in technology have generated a new river of scholarly papers on the subject, many so intricately technical that narrow specialty fields have developed that are accessible only to the initiated few. The ice ages have thereby become the most exhaustively studied of all geological epochs, and the mountain of literature on the subject counts as one of the citadels of modern science.

But close examination reveals weaknesses in this fortress nevertheless. One finds that common sense must be stretched alarmingly at times in order to accommodate the evidence, and some features are so starkly contradictory that they have to be set aside and ignored altogether. In any other field of inquiry such profound problems would condemn a theory outright, but the ice age premise is a special case because it simply has no obvious alternative. However, one not-so-obvious alternative was proposed by Ignatius Donnelly who thought to trace the relics in question to a collision between the Earth and a comet [30]. But despite impressive supporting evidence his views were not taken seriously—in part because he erred grievously in his understanding of comets.

Nevertheless, he did expose many serious flaws in the ice age premise that remain valid flaws even today. Indeed, among his many general arguments three stand out as especially persuasive evidence that some alternative to the ice age picture must exist; it has only to be found. Obviously if a mistake was made in interpreting those old residues then it was made at the very inception of the theory, so it can be corrected in that same early context. Since we will be starting over from the beginning we shall not be obliged to sift through the mountain of learned papers that came later; only the most superficial properties of the residues will have to be considered. We begin, then, with a brief survey of the materials in question.

They are variously known as till, boulder clay and drift, depending on their location and predominant characteristics. Till is a very old term that was first used in Scotland to describe the tough, hardpan clay so prevalent there. It is an exceptionally compact red clay that can be used for making bricks, and it is usually the first layer of deposit above the bed rock. Stones of various sizes are distributed randomly through it which often bear deep grooves and striations caused by an abrasive process of some kind. These striations tend to run parallel to the longest dimension of the stones, although this is not invariably the case. The proportion of the total mass contributed by the stones is widely variable. Flint describes the material in this way [33; p.154]:

> "... It may consist principally of clay particles or principally of boulders or any combination of these and intermediate sizes. Grain-size analysis of a sample of till is accomplished by disaggregation followed by passing the particles through a series of sieves of diminishing mesh. For analysis of the clay-

Ice-age Residues?

sized particles a more complex technology must be used. The particles stopped by each sieve are then weighed and calculated as a percentage by weight, of the sample as a whole. ... More commonly it is expressed as a cumulative curve ... in which grain-size is plotted against percentages that are smaller than the stated diameter.

" The curves for most tills are bimodal or even multimodal. One cannot confidently explain such asymmetry merely by saying that some sizes were flushed away before deposition."

As we found in the last chapter, and as Flint explains here, it is exceedingly difficult to explain how any single natural process of selection or grinding can yield a bimodal grain-size distribution. Such a distribution requires at least two processes. Either it results from the intimate mixing of the products from two different processes, or it requires the selective removal of a given size range from an otherwise normal distribution.

But it is no easier to imagine either of these situations prevailing in a glacial environment than at Papago Park. A glacier might pass over different rocks locally which would abrade in their own way, but such local variations should average out in the end product to yield the statistically expected simple distribution of grain sizes. The curve that Flint described above is, of course, only a different way of reporting the same data as given in a histogram such as Figure 5. In his representation a bimodal representation yields a curve with three inflection points instead of only one.

Returning to our summary, boulder clay is closely akin to the till and, in fact, may be merely a local variation of that same material since the two forms are sometimes seen to grade smoothly into each other. In this form the included stones tend to be larger but not always striated, and the clay is not as tough. The stones within this deposit can attain to enormous sizes—tens of thousands of cubic feet in some cases, and they are often isolated from any possible source by upwards of a hundred miles. Even as the till, the boulder clay is typically unstratified—namely, it is not arranged in layers or beds.

Drift is a more general term that includes both of the above as well as other similar unsorted and unstratified massive deposits that overlie the topmost layer of bed rock. The term brings to mind alien material that has been carried to its present location through the agency of moving ice. It consists of gravels, sands and clays mixed in all proportions, and the deposits commonly measure several hundred feet in thickness. It is noteworthy and significant that the drift is, with only rare exception, entirely fossil free. Examples of these three classes of material are found covering most of the northern United States and Canada as well as northern Europe, but it is interesting that none are to be found in all of China or Siberia.

The first of Donnelly's key points is that till is such a remarkable mixture that, taken alone, without any other clues, one would be hard pressed to imagine how it could have been formed. Certainly this intimate mixing of all possible grain sizes removes the agency of water from consideration, and that leaves so puzzling a mystery that presumably it can have at most only one solution. That is, whatever mechanism was responsible for forming the till, that same mechanism must have given rise to all occurrances of it—wherever found.

Inexplicably, from the glacial point of view, till is not confined to the northern regions but is found in the tropics as well; in fact it even covers a large part of Brazil! In order to contend with this problem, Flint proposed [33;p.154] that a new generic name, "diamicton", should be coined to include all "nonsorted terrigenous sediments containing a wide range of particle sizes, regardless of genesis." Then he would reserve the name "till" for only those diamictons that were of glacial origin. Thus would he define away the Brazilian till. But Louis Agassiz, "grandfather" of the glacial theory, had no doubt that the "ferrugenous clay with pebbles" of Brazil was indeed of

Ice-age Residues?

glacial origin [30;p.37].

Whether of glacial origin or not, assuredly it must be traced to the same source as the till in Scotland and elsewhere. But to theorize glaciers from pole to equator is to theorize the extinction of all life on earth. Evidently, then, the till must be traced to some other origin than glaciers.

Secondly, Donnelly cited James Geikie's observation that the base rock upon which lies the till in Scotland is often broken and shattered. Here is how that Scottish geologist described these rocks [35;p.24]:

> " The pavement of rock below till is frequently smoothed, polished, and striated, more especially if the rock be hard and fine grained; softer rocks, like sandstone and shale, and highly jointed rocks like greywacke, and many igneous masses, often show a broken surface beneath the till."

This is a remarkable set of circumstances which is worthy of careful attention. Donnelly deduced from this smashed appearance [30;p.52] that some violent catastrophe preceded the laying down of the till—one that shattered the basement upon which the clay would soon lie. He urged that it was this explosive cataclysm that also gave rise to those striations in the base rock.

But this shattered condition of the basement rock also proves that no glacier ever moved over it. This conclusion follows easily from the manner by which moving ice performs its abrasive action. Namely, in the deeper regions of glaciers ice under pressure flows into crevices in the rock, and the shear stress exerted by the moving ice forcibly pulls pieces out. These "quarried" pieces become the tools which then score the basement as they are carried along with the ice, and they end up finally in the terminal moraine. Therefore if a glacier had actually generated those striations, then it would also have quarried those shattered pieces of basement rock and carried them away.

Finally, Donnelly pointed out that the glacial theory offers no possible accounting for the clay that forms such an important component of the drift. Let us keep in mind that clay is not merely finely ground rock—sometimes called "rock flour"; it is a mineral in its own right and is derivable by weathering primarily from feldspar. This is a white, crystalline mineral, and clay derived from pure feldspar is white also—or nearly so. The red coloring present in most clays derives from a certain admixture of iron oxide.

Donnelly noted that feldspar most commonly occurs as a constituent of granite, which also contains quartz and the iron-bearing minerals, mica and hornblende. He also urged that granite is not commonly exposed at the earth's surface, being normally buried under thick layers of sedimentary rock. So the problem is one of simple chemistry. Namely, it is inconceivable that the vast expanses of clay observed in the till could have been derived from the relatively rare exposures of granite—and even less conceivable that the glaciers could have picked out the feldspar and mica from the granite in order to produce the pure clay—and less likely yet that they could have picked out only the feldspar so as to produce the nearly white clay that is found west of the Mississippi.

Comyns Beaumont was another articulate opponent of the ice age point of view. In addition to the points noted above he urged that an actual climatic phenomenon would have affected both the northern and southern hemispheres equally [9;p.33], and yet there are no equivalent deposits south of the equator—except in Brazil. Moreover, he observed that the drift is not even distributed symmetrically around the north pole; most of Siberia, Alaska and Canada being free of these relics altogether.

Each of these points is compelling by itself, but taken together they must be considered definitive. Add to all of these the strange bimodal character of the grain-size distribution and one can say with confidence that the till and associated drift

Ice-age Residues?

deposits were not formed by any kind of glacial action. Those early geologists would not have thought so either if any other alternative had been apparent to them. But because no alternative could be imagined succeeding generations of geologists have been building upon that enforced option ever since.

With that point firmly settled we can move forward and try to discover just where the drift did come from. In actual fact, with a mind completely open to the force of reason, one could probably deduce the origin of the till from what we already know, but the surprising resolution to the riddle will be all the more convincing if we pause to gather a little more data in advance. Of all the descriptions of drift deposits that have been published in the geological literature none can be more revealing than N.H. Winchell's firsthand account of one notable occurance in southern Minnesota. Let us attend to his very words as he tells about it [83; p.104]:

> " The most important fact in connection with the drift of these counties [Rock and Pipestone counties] is a gradual transition, from north to south, from drift clay, with stones and boulders, to loam clay that has all the characters of the well-known loess-loam of the Missouri valley. The northern part of Pipestone county lies not far from the Coteau du Prairie, which is a vast glacial moraine of drift materials, and is even affected somewhat in its contour by the westward decline of the Coteau to the prairie level. It is as characteristically a hardpan clay—the main mass of the drift, in this part of Pipestone county—as in any part of Minnesota. In travelling southward there is a gradual superficial change in all its characters. This change pervades at first but a small thickness of the deposit but by degrees involves the drift to a depth of 20 feet. At first there is a diminution in the number of visible boulders; then a smoothness in the creek bluffs; then a gravelly clay on the surface, fine and close; then a closeness in the prairie soil; then, in digging wells a few limey concretions are seen mingled with the gravelstones, and at last a fine crumbling loam clay that cannot

be distinguished from the loess-loam, which extends to Sioux City in Iowa, and is there known as the loess-loam of the Missouri valley and has a thickness of several hundred feet. Wells dug in the southwestern part of Rock county demonstrate also a similar perpendicular transition from loam to drift clay, the former being true loess-loam and the latter true hardpan, or boulder clay. This appears like rank heterodoxy, but it is not a matter of opinion or theory. It is a result of actual observation. The writer was as much surprised to find it as others will be to read it, and it appears almost inexplicable."

Just what is this loess-loam and why should Winchell have been so surprised at that gradual transition between the two materials? The reason is easily given, and his observation is so profoundly significant that we shall do well to examine the loess in some detail. Our study of this material will consume the remainder of this chapter and part of the next. At the end of this inquiry the reader should feel very confident of his understanding, and the reason for Winchell's surprise at that unexpected uniform transition will become plainly evident.

YELLOW EARTH

For millions of people, upon four continents, loess is the very dirt beneath their feet. Nevertheless it is a very special kind of dirt. Its most obvious features are a silty texture and a yellowish color, but it has many other special features which we shall consider presently. It is localized into several broad areas around the globe; in the United States the deposit begins in central Nebraska and extends south and east to cover much of Kansas, Iowa and Missouri. It thins further to the north and east, but it continues to the south along the valley of the Mississippi and shows up especially well near Vicksburg and Natchez. Extensive deposits are also found in South America, notably in Argentina and Uraguay. Likewise, it is found locally in northern Europe and more broadly in central Asia, but by far

Ice-age Residues?

the greatest deposits of all are found in China where the loess covers nearly a million square miles! In fact, to the Chinese yellow is the color of mother earth. The Yellow River derives its name from its burden of eroded loess, and the Yellow Sea is similarly colored by this same material.

The native, undisturbed loess contains a small proportion of calcium carbonate (that is, calcite, the predominate mineral in limestone) which acts as a cement and binds the individual grains together. This gives the material a firmness

PLATE 25: *Exposed loess along Route 275 at Council Bluffs, Iowa showing well-defined vertical cleavage.*

that silt alone could never have. So firm is it in fact that in China very acceptable living quarters have been carved out of the loess deposits. But it is less firm when moist. Although calcite is not readily soluable it does dissolve slightly, so water frees the grains to some extent. Running water, then, removes the calcite entirely and quickly washes the silt away. As a consequence, streams of water easily erode through the loess and form their beds upon the firmer substrate below.

It is also an uncontested fact that the loess buries the pre-existing landscape like a blanket, hill and valley alike. Its thickness varies upwards from a hundred feet and down to the point where it can no longer be distinguished, and wherever

PLATE 26: *An old photograph of loess bluffs in China. Reproduced from Plate XXIV of Reference 82.*

Ice-age Residues?

found it presents the appearance of a massive, uniform deposit. Plate 25 shows an exposure of the loess along Route 275 in Council Bluffs, Iowa. One can also make out another typical and extremely significant property of the material in this view; namely, it has a tendency to cleave along vertical planes so as to form a bluff. this is quite general, and in fact the city of Council Bluffs derives its name from the loess bluffs which extend for miles here along the valley of the Missouri River. Plate 26 is a photograph made in China many years ago that shows this surprising feature very clearly. The bluffs retain their form for many years, and when undermined by further erosion a section will collapse to yield a vertical bluff formation just as it was before.

This surprising tendency to cleave along vertical planes is due to the fact that the undisturbed loess, wherever found, is perforated with countless capillary tubules which are oriented mainly in a vertical direction. And here we come upon the first solid indication that his material has a far from an ordinary history. In fact, these tubules are so unexpected that perhaps we should hear an eyewitness tell about them. The following description by Baron Ferdinand von Richthofen* was translated from the original German by the authors of Reference 82, and it reads as follows [p.183]:

> " On every bit of loess, even the smallest, one may recognize a certain texture, which consists in that the earth is traversed by long-drawn-out tubes, which are in part extraordinarily fine and in part somewhat coarser; which branch downward after the manner of fine rootlets and generally are coated with a thin white crust of carbonate of lime. If one

---

* *Modern geological studies of the loess actually began with Baron von Richthofen who published a treatise on his research in China in 1877. This now rare work was the ultimate source of the passage quoted here. Its author was an uncle of Manfried von Richthofen—the "Red Baron", famed German flying ace during the first World War.*

examines the loess in place one sees that most of these little channels are nearly vertical, yet branch at an acute angle and downwards, whereby an incomplete parallel structure is maintained. If one is looking at a loose piece, but not exactly at the surface of parallel fracture, one sees the ends of the little tubes which occasion an appearance of minute holes. But, apart from these definitely bounded elongated spaces, the earth between them has a loose porous structure and does not possess that close texture which is peculiar to other kinds of earth; for example, the clays, potter's clays, and loams."

Notice that the tubes actually constitute a key identifying texture of the material since they are to be found "On every bit of loess, even the smallest." Notice also that the baron described these tubes as "branching downwards after the manner of fine rootlets." Because of this branching the most commonly cited explanation would have the tubes to be holes left by the roots of small plants which grew upon the silt while the formation was developing. Those who support this idea point to the fact that tubes can sometimes be found with a root residue still inside. Without a doubt the presence of a dried root within a tube is persuasive evidence that that particular hole was formed by a root, but it would seem to argue just as strongly that the empty tubes must have some other origin. A more plausible accounting for the uniform presence of these tubules throughout the entire body of the loess appears to be required, and one will be offered in its proper place.

In his description above Baron Richthofen also mentioned another very significant property of this material; namely, he pointed out that the native loess is extremely porous, even apart from the tubes. In fact, according to Flint the porosity generally exceeds 50 per cent [33;p.252]; that is, more than half of its bulk is entirely vacant! One contributing reason for this strangely porous texture may be found in the unusually uniform size of the particles which make up the material. Most of the grains are in the size range of a fine silt, with the smaller

Ice-age Residues?

PLATE 27: *A scene from China showing deep erosion of a roadway into the loess. (From Plate XXIV of Reference 82)*

clay-sized particles being rare and sand grains being almost nonexistent. Of course, tight packing requires a broad range of particle sizes in order that the voids between the larger grains can be filled by smaller and ever smaller particles to form a compact whole. Because the component particles are so nearly the same size, that kind of packing is impossible in the loess, but even this cannot be the full reason for its great porosity. For when the binding cement is washed away and the silt is disturbed or redistributed the grains do settle together somewhat, the packing becomes more dense, and the material then resembles loess only in its color. In this condition it is usually called loess-loam, and it forms the basis of an extremely fertile soil—as farmers in Iowa are quick to boast.

The uniform size of the grains making up the loess, and the resulting inability of the particles to pack tightly together, makes the silt extremely vulnerable to erosion by the wind whenever it is once disturbed. When the binding cement is lost, its strength also vanishes. Plate 27 illustrates one interesting consequence of this vulnerability to erosion that was once seen fairly typically in China. Here the road has worn deeply into the loess as the newly disturbed fine silt, leached of its binding cement, has continually blown away.

Scattered throughout the loess there normally occur peculiar nodules such as are shown in Plates 28 and 29. The Chinese call them "loess ginger" and the Germans, "loess dolls". They are especially rich in calcium carbonate, some being as white as chalk while others, containing less calcite are much the same color as the loess itself—as is the specimen in Plate 29. The peculiar surface texture of this one suggests that the nodules have an interesting story of their own to tell, but let us set them aside for the present and continue with this first introduction to the topic.

Although his original treatise is rare today, Baron von Richthofen reviewed his observations a few years later in a more accessible journal so we can hear this scholar describe,

Ice-age Residues?

from his own firsthand experience, some of the properties that have just been discussed. His English is patterned somewhat after his native German so one must be especially attentive in order to follow his meaning, but let us hear his own words nevertheless [68; p.295] :

" Any theory which undertakes to deal with the problem of the origin of the Loess must give a valid explanation of the following characteristic peculiarities of it, viz.:

" 1st. The petrographical, stratigraphical, and faunistic difference of the Loess from all accumulations of inorganic matter which have been deposited previously and subsequently

PLATE 28: *Specimens of the characteristic limey nodules found scattered throughout the loess.*

to its formation, and are preserved to this day.

" 2nd. The nearly perfect homogenousness of composition and structure, which the Loess preserves throughout all the regions in which it is found on the continents of Europe and Asia; it offers in this respect a remarkable contrast to all sediments proved to be deposited from water within the last geological epochs, excepting those of the deep sea, which are here out of the question.

" 3rd. The independence of the distribution of the Loess from the amount of altitude above sea-level. In China it ranges from a few feet to about 8000 feet above the sea, and farther west it rises probably to much greater altitudes. In

PLATE 29: *Another example of loessian nodule, this one darker in color and with striking surface features.*

Ice-age Residues?

Europe it is known at all elevations up to about 5000 feet, at which it occurs in the Carpathians.

"   4th. The peculiar shape of every large body of Loess, as it is recognized where erosion has cut gorges through it down to the underlying ground without obliterating the original features of the deposit. These are different according to the hilly or level character of the subjacent ground. In hilly regions the Loess, if little developed, fills up depressions between every pair of lower ridges, and in each of them presents a concave surface; but where it attains greater thickness, it spreads over the lower hills, and conceals the inequalities of the ground. Its concave surface extends then over the entire area separating two higher ranges, in such a manner as to make the line of profile resemble the curve that would be produced by a rope stretched loosely between the two ranges. ...

"   5th. The composition of pure Loess, which is the same from whatever region specimens may be taken, extremely fine particles of hydrated silicate of alumina being the largely prevailing ingredient, while there is always present an admixture of small grains of quartz and fine laminae of mica. It contains, besides, carbonate of lime, the segregation of which gives origin to the well-known concretions common to all deposits of Loess, and is always impregnated with alkaline salts. A yellow colouring matter caused by a ferruginous substance is never wanting.

"   6th. The almost exclusive occurrence of angular grains of quartz in the pure kinds of Loess."

Here is eyewitness testimony to one of the key features of loess; it spreads like a blanket over the former landscape, irrespective of elevation. Richthofen points this out directly as his 3rd property and then he amplifies upon it in the 4th. In fact, his words here might be used to describe a heavy snowfall just as well. The baron concluded his observations as follows:

"   There is but one great class of agencies which can be called in aid for explaining the covering of hundreds of thousands of square miles, in little interrupted continuity, and almost irrespective of altitude, with a perfectly homogeneous soil. It is those which are founded in the energy of the motions of the atmospheric ocean which bathes alike plains and hill-tops . ...."

His language may be a bit flowery, but the central idea is plain enough; the silt must have filtered down through the atmosphere from above. It could not have been deposited by, or through water in any form—neither in seas, lakes, nor rivers.

Thus was born the so-called "aeolian theory" of loess deposition. One can appreciate the dismay that geologists must have felt on being confronted with such a confounding list of properties to accommodate within the framework of a single picture—and especially, as Baron Richthofen urged, the *same picture* wherever it is found. And on top of everything else is the added fact that this airborne material grades smoothly into dense, rocky clay which is supposed to be the residue of glaciers here in North America. The reason for Winchell's surprise at this discovery is now painfully obvious. One must marvel that the glacial theory survived it at all, but fashionable opinion is as an irresistable tide; it has a life of its own.

However, let us resolve to rise above the tide and search out the truth for ourselves. Continuing with our studies, then, Baron Richthofen suggested that the silt making up the loess in China had its origin in the Gobi Desert to the north. He thought the material was a product of desert erosion which the wind had carried southward; presumably it sprinkled down upon the land as the wind lost its force. This idea was generalized somewhat in later years to accommodate it to the particular locations where loess is found. For example, there are no vast deserts in Europe to which the European silt could be traced, but there are extensive drift deposits so these are usually offered as the source instead.

One common theory pictures vast flood plains formed by

Ice-age Residues?

outwash from the melting glaciers. Then, when the water abated, and when the mixture of mud, silt and sand had dried, the wind is thought to have picked up the finer particles and blown them away. The fine silts were presumably redeposited close at hand to form the loess accumulation, while the finest particles were distributed over a broader area and were effectively lost.

But this picture is beset with very serious problems. For one thing, it does not account for the calcite present in the loess which binds the particles together. If the supposed residue on that flood plain had contained grains of calcite (which the flood waters did not dissolve away), then, as the muddy mixture dried, the particles would have been strongly cemented into a rigid mass which the wind could not have disturbed. On the other hand, if that residue did not contain calcite then where did the loess obtain its uniform supply? Although the source of the silt assumed by this theory is altogether different from that pictured by Richthofen for the loess in China, the wind is still assigned the key role in its deposition. Supposedly it picked up the silt from one place and dropped it down in another so as to form that typical blanket-like covering of the former landscape.

In the United States we have both the western deserts and the drift deposits in the north, and each has its defenders as the source for the silt. But it might be well to recall again the essential conditions that Richthofen insisted must be satisfied by any plausible theory of the loess. Among other things he urged that any valid theory should account for the fact that the properties are the same regardless of the locality where found. It is not easily possible to understand how this same end product (complete with tubes and nodules) could have been derived from such varied source material and under such widely different local conditions.

But as if that were not enough there remain still other difficulties with the prevailing views which, although seldom emphasized today, once sparked a lively opposition to the developing trend of thought. It is worth our while to examine these problems thoughtfully for they will strengthen our conviction that present theories of the loess simply cannot be correct. The trail leads eventually to some very startling conclusions so one needs all the confidence in his understanding that he can muster if he is to face up to them squarely when the need arises. We shall learn about these problems best by attending to some of that early discussion as it came from the pens of those who opposed the current views most vigorously. Here, for example, is how H. H. Howorth argued one point more than a hundred years ago [49; p.348] :

> "... the Loess for the most part is completely unstratified. Occasionally, especially in America, there are local areas where a kind of stratification occurs, but these are very local, and I shall return to them presently. This absence of stratification I quoted myself as a proof that the loess is neither of marine, lacustrine, nor fluviatile origin [that is,were formed neither in seas, lakes nor rivers]. It is assuredly equally a proof that it is not due to gradual accumulation by the wind. Dunes accumulated by the wind are so easy to study that we have no difficulty in finding materials, and assuredly they present quite a different structure to Loess. Deposits made by wind, especially when made as Baron Richthofen suggests, in dry seasons alternating with wet ones, have a laminar structure corresponding to the series of layers deposited, just as deposits made by water have. Nor should we find homogeneous masses several hundred feet thick with the same structure and the same contents as the result of such a series of seasonal deposits. These masses, to my mind, bear, on the contrary, unmistakable evidence in their very structure of having been deposited by one great effort, and under one set of conditions. ... "

Now here are several interesting observations. For one thing, Howorth points out that the detailed structure of the deposit bears no similarity, in general, to sand dunes which

Ice-age Residues?

have been blown and redistributed by the wind. From this fact one must conclude that the calcite was part of the original complement of grains that settled down to form the loess; it was not added afterwards in some unknown way as some have suggested. Evidently the mass was "frozen" into place by the calcite the first time it was moistened by the rain. The small crystals would dissolve slightly; then they would grow again as the water evaporated, and in growing anew they would cement the contact between neighboring grains. If the calcite had not been present in the mixture from the very beginning then dune features would have developed while the silt was still free to be moved about by the wind.

Notice also that Howorth objected to the idea of *gradual* accumulation by the wind. And how could it have been anything but gradual if the silt had been derived from erosion in the Gobi Desert? Erosion is a very slow and gradual process, after all. In his paper he went on to argue that if the material had been deposited gradually then plants of all kinds would have grown upon it—large plants that would later have been buried in it. But the loess is far different from what one would expect of such a gradual accumulation. As he says above, it bears unmistakable evidence of having been deposited by one great effort and under one set of conditions.

L. S. Berg was another bitter adversary of the prevailing aeolian theory, and he also waged a long fight against it. Here is how this staunch opponent phrased one of his several objections [11; p.134]:

" ... it is absolutely incomprehensible, why the wind should drive sediment of only that texture which is characteristic of loess. The wind, according to its velocity, can carry either coarser or finer particles, but why it should give preference to particles of 0.01 to 0.05 mm. in diameter, has never been explained by any follower of the aeolian theory. Typical loess being characterized by the predominance of particles of the above mentioned diameter both in Europe, Asia and America, we should have been forced to conclude that the wind had everywhere the same velocity. Moreover, the wind would have to blow in the same direction and with the same velocity during tens of thousands of years. Otherwise the sediment that had settled down could never be so uniform. As in actual fact winds blow with varying force and from varying directions, it is evident that had aeolian loess existed it would have been a mixture of particles of the greatest variety of coarseness, the texture of loess in neighboring areas at the same time being very diversified. ..."

As one considers Berg's argument with the enormous extent and uniformity of the loess deposits in mind ("hundreds of thousands of square miles in little interrupted continuity ") its force becomes irresistible. Furthermore, we know very well that the wind carries dust easily and deposits it everywhere, yet these finer particles are notably scarce in the loess.

But problems with the aeolian theory only multiply as one probes deeper, for as it happens gravel and even pebbles are sometimes to be found in the loess! Are pebbles, then, carried on the wind while sand and dust are not? The following example of pebbles in the loess in China is somewhat difficult to follow because the place names are unfamiliar, but one can hardly fail to get the point nevertheless. The authors gave all of their measurements in both feet and meters, but these latter have been omitted for the sake of clarity [82; p195]:

" Between Ling-shi-hien and P'ing-yang-fu the Huang-t'u [loess] covers the uplands up to more that 1,500 feet ... above the river. Underlying strata of the Shan-si coal-measures are exposed in many ravines, but the slopes are buried in fine silt to depths that range from 100 feet ... to possibly 300 feet. ... The deposit is thickest next to the valleys, and is there interbedded with layers of coarse wash. Thus, north of Yon-yi- ssi, at an altitude of 700 feet ... above the town, there is a 30-foot ... bed of pebbles up to 5 centimeters in diameter. The bulk of the

Ice-age Residues?

material appears to be loess, but in sections seen along the road the true constitution is obscured by a coat of dust, and coarser sand and gravels may be present in larger proportions than one expects. Vertical cleavage is everywhere characteristic. ..."

The authors suggested that the interbedded material was "wash" because it occurred in layers; they thought it had been laid down by running water. But running water quickly carries the loess away so it could not have been wash in any sense of the word. And of course, the mixture they described could not have been lifted up by the wind and carried from some other source. This is easily as profound a riddle as Winchell observed in the smooth drift-to-loess transition in Minnesota.

But probably the most remarkable example of "pebbles in the loess" ever to come to this writer's attention was reported by Skertchly and Kingsmill who described a feature of the loess in China. Let us read an extract from their paper, fixing our attention on the "old river gravels" and being not distracted by the perhaps unfamiliar term "Carboniferous" [74; p.243]:

" The old river-gravels of Shantung form a very interesting set of deposits ... they constitute a conglomerate of lime-stone pebbles in a calcareous cement, the component fragments varying from fine gravel-stones to heavy shingle. ...
" The conglomerate is exceeding hard, and hence has resisted the denudation that swept away the looser material which doubtless originally accompanied it. The beds often stand out as bosses and banks upon the loess-plain, just as, and for similar reasons, many of the ancient gravels of Cambridgeshire and Norfolk rise above the Chalk. So eminently calcareous are they, that in places where no limestone is locally available, as at Chow Ts'un ... , they are used for limeburning. From their very nature they are sometimes only a few feet in length, sometimes continuous for hundreds of yards, and from 8 to 30 yards in width. So compact is the mass that fragments 6 or 7 feet wide and 20 to 30 feet long are often seen projecting like shelves from the loess, and at one place we rode through a chasm 15 feet wide

spanned by one of these masses hardly a foot thick. It is frequently used as rough building-stone for retaining-walls.

" In the great Carboniferous Limestone district which flanks the sacred Tai Shan range ... , these relics of prehistoric rivers are still to be seen clinging to the valley-sides. They can be traced at intervals from the hills out onto the loess-plain, and not unfrequently five or six may be seen at different levels on a valley-side. About 80 per cent. of their courses are quite independent of the present drainage-system, and beds sometimes cross each other. Often, too, the roads, which from centuries of traffic have become worn below the surface of the soil, climb laboriously over one side of these old courses, and as abruptly descend on the other, ...

" In many respects the old river-beds differ widely from those of streams now traversing the same district. The present courses are, where paved with gravel, worn down to the bedrock, while the old beds frequently lie from 50 to 100 feet above it, the gravel often resting on a level bed of rearranged loess. Then, too, in the recent river-beds, wherever the gravel is made of limestone-pebbles there is no tendency to cement into conglomerate. ..."

This report is so amazing that it is worth reading for a second time. Could those great limey slabs actually be residues of an old river? One would hardly think so because no river can flow upon the loess. Running water quickly wears through the silt and washes it away as the authors were quick to point out in their final paragraph above. Furthermore, they point out that 80 per cent of the courses of those "old rivers" were independent of the present drainage system—in other words, the slabs were scattered about at random. Certainly they had nothing to do with old rivers. Here is a problem of heroic proportions indeed, but can it be distinguished from that of the loess as a whole? Is the mystery of these slabs any more profound than that of the ever-present tubules or the fact that the silt grades smoothly into boulder clay? Clearly we have but one riddle to

Ice-age Residues?

resolve here, of which the pebbles, the great limey slabs and the tubules are merely separate parts. A true understanding of the loess must provide for all of these peculiarities in the same context at one and the same time.

Now for the sake of completeness we should back up here and attend to a few peripheral points that bear on this problem of origins. Recall that Howorth referred to "a type of stratification" present in the American loess that slightly qualified his argument against seasonal deposition by the wind. This stratum-like appearance can sometimes be noticed in bluffs or where roads have been cut through the loess, and it shows up as changes in color, or changes in the intensity of color, of the material. These bands of color variation are often associated with discernible changes in composition as well.

According to the current view of loess deposition these darker bands constitute "fossil soils", and they are called "paleosols" on that account. That is, each of them is thought to have once been a humus-laden layer upon the surface which formed when the deposition of silt was temporarily interrupted. Of course, the loess is now thought to have been deposited over many thousands of years, and that would allow ample time for such soils to have developed. But if instead the silt was deposited quickly—"by one great effort, and under one set of conditions", as Howorth insisted, then there would have been no chance whatever for soils to have developed during supposed intermediate stages. It is therefore important to know if those darker bands actually were soils at one time, or whether this is merely a guess having no actual foundation. Here is Berg again, reminding his colleagues what chemical analysis reveals about one aspect of the question [11; p.135]:

> " The followers of the aeolian theory affirm that loess is formed not in deserts, whence loess dust is blown off by winds, but on the periphery of deserts, in steppes, where vegetation contributes to the accumulation of loess. Thus, according to this notion, the whole profile of loess had to pass through the stages

« 143 »

of soil formation, namely of chernozem, or, at least, of the chestnut type of soils. But in that case a considerable quantity of humus should be present in loess, what, as is well known, does not take place; the content of humus in normal loess is manifested in tenths or hundredths per cent, and sometimes it runs down to naught. One might say that humus had been present once and had subsequently decomposed. But loess beds, as we know, are generally inter-stratified with one or occasionally several fossil humus horizons in which humus is unaltered; although little remains in these ancient soils, from 0.3 to 1.1 per cent on average, but a humus horizon is always distinct. Thus, if loess had been formed by steppe vegetation being buried under an accumulation dust, the entire profile of loess should exhibit a semblance of a humus horizon. But such in fact is not the case. Therefore loess could not have been deposited in steppes."

So Berg went along with the idea that the colored bands were fossil soils, but he noted that their actual humus content was very small. He thought that humus had once been plentiful in them but that most of it had washed away during the succeeding years. But might it have been the other way around instead? Plate 30 shows another exposure of the loess at Council Bluffs which plainly shows the sharp contrast between the rich soil layer that has developed at the surface and the natural loess below. One of those darker bands is also evident somewhat below the surface. It is surely significant that the material within this darker band exhibits the typical vertical cleavage just as plainly as does the material above and below. Therefore the tubules are well preserved in this region too, so the material could never have been soil in this sense of the word; its organic component must have some other origin.

Perhaps it might be traced most reasonably to the rich soil layer at the top. Some of the soluable humus components at the surface would naturally filter down into the underlying layers, and if the composition of the loess varies slightly in the

Ice-age Residues?

colored bands then the filtering capacity would probably vary as well. In that case the colored bands need not have lost a once significant humus content as Berg imagined; more likely those traces which they possess were acquired by seepage and filtration from the soil layer at the top.

Let us consider those strange tubules again in light of the information which Berg has just given. If they had been formed by grass or other small plants growing upon the accumulating silt then organic residues of those plants should still remain, and chemical analysis ought to reveal them as a humus residue, for it is difficult to imagine that such humus could have washed away while the calcite binder still remains intact. Furthermore,

PLATE 30: *A view of exposed loess near Council Bluffs, Iowa showing a well-defined layer of soil upon the surface.*

if the silt had deposited slowly then larger plants would have grown upon it as well, and their residues also ought to be found entombed within it. Even if those plants had decomposed entirely, and the organic products of decay had washed away completely, then one should still expect to see the casts of those larger plants preserved as large tubes and holes even as the small tubules are perfectly well preserved. But nothing of this kind is found.

Baron Richthofen, we recall, observed that the loess covered the former landscape like a blanket, irrespective of elevation, and he drew the unavoidable conclusion that the silt must have filtered down through the atmosphere from above. That much is undeniably true, but the aeolian theory as it stands today falls grievously short in every other respect. It does not explain how the loess could grade smoothly into the boulder clay nor how pebbles could sometimes be found within it. Likewise it does not account for the unlikely porosity of this material nor does it explain how this same peculiar mixture could be found as massive, localized deposits on four widely separated continents. And then finally, it does not explain the absence of humus in the loess—along with the casts of plants that would have grown upon it while it developed. At this point one ought to be able to declare in perfect confidence that the colored bands are not the residues of former soils, the capillary tubules are not root holes, the silt did not deposit slowly over great periods of time, and finally, and most importantly, it was not picked up by the wind and transported from some former location. Without a doubt we must look elsewhere for understanding, but what are the alternatives?

One interesting possibility was suggested in 1920 by K. Keilhack. He reasoned that if the loess did actually come from the drift formations, as was commonly thought in the case of the European loess (and still is) then, since those residues contain great quantities of sand in addition to silt, there should also be found residual sand deposits comparable in size to the

Ice-age Residues?

loess itself. That is, Keilhack looked for the sand that remained after the silt had been removed by the wind, but he did not find it. Therefore he eliminated the drift residues as a possible source for the silt, and offered a thought-provoking alternative in their place. The following excerpt is a close translation from the original German so some extra care is required in order to follow his reasoning. Here, then, is the way Keilhack analyzed the loess problem [52; p.158]:

" ... Calcite and quartz, both important ingredients of the loess, must have been taken by the wind from entirely different places on the earth, for we know of no rock in which they occur together in the proportions and the grain size of the loess since we have had to set aside the glacial formations from consideration. Where, however, and in what way was this astonishingly uniform mixing brought about as we see it in the loess today—a mixing of these two so different constituents that shows little variation over the whole earth. Can we explain the uniformity in composition in Europe and Asia, in North and South America in any other way than that all the loess masses of these four vast regions have drawn their material from the same large mixing bowl? Must not then the mixing have taken place at a very considerable altitude? The size of the loess particles argues against this, however; they are too large to have remained suspended in the high atmospheric layers for a day, to say nothing of years or hundreds of years. So many questions, so many entirely unresolved riddles!

" Starting from the recognition, or the probability, that the entire terrestrial loess is a uniform mixture, and assuming a common reservoir out of which the deposition took place, it is only a step to raise the question whether, in that case, an extraterrestrial, a cosmical origin for the loess can be entirely excluded? Here, of course, the astronomers have the last word. However, I would not hesitate to suggest it is through such assumptions, bold as they may seem today, that some of the riddles posed above do find a satisfactory answer, ..."

But that final word has always come back very clearly in the negative, for according to our customary manner of thinking, the terms "extraterrestrial" and "cosmical" mean the same thing; they both refer to outer space—the interplanetary domain. And there can be no doubt that if enough material were to fall from outer space upon the earth to form the massive, localized accumulation characteristic of the loess then it would necessarily fall as a meteorite falls, with high velocity and great energy; the atmosphere would not be able to absorb the energy of such a huge falling mass. From a cosmical encounter on this scale, then, an exceedingly violent explosion would normally be expected when the material struck the Earth, and a gigantic crater would be formed, with debris piled high around the rim—but of course nothing of the kind is found.

Casting about in The Laboratory for an agreeable "scientific explanation" for this strange silt, with its pebbles, conglomerate slabs and other remarkable properties, one can only throw up his hands in despair; certainly no understanding is remotely possible within the confines of this artificially limited space. In desperation, then, let us dare to step outside of The Laboratory and meet Nature face to face.

Chapter 7:

OPENING THE DOOR

MODERN GEOLOGY was born at a time when the Biblical view of origins held sway over most of Europe. Its free-thinking founders chafed at the rigors of an imposed orthodoxy and wished to establish earth studies on a par with the other natural sciences. They were, however, at a profound disadvantage. Physics and chemistry had ascended to the status of sciences at last when great masses of existing empirical data could be distilled into a few general principles. But no corresponding mass of detailed knowledge of the earth's structure even existed at that time so there could be no careful examination of broadly based evidence—and even less a distillation of it. In their eagerness to break free of religious dogmatism the fathers of modern geology devised as their fundamental principle not a distillation of existing factual data but an aspiration instead. This aspiration is commonly known today as the Principle of Uniformity.

This Principle is variously expressed, but perhaps most simply by the dictum, "The present is the key to the past". Proposed more than two hundred years ago by James Hutton of Edinburgh it may be explained as follows [37; p.18]:

" ... Applied specifically this means that rocks formed long ago at the earth's surface may be understood and explained in accordance with physical processes now operating.

" The Uniformitarian Principle assumes that the physical laws now operating have always operated throughout the geologic past. It assumes, for example, that in the geologic past, just as today, water always flowed downhill, collected into streams, and carried loads of mud and silt to the sea. It assumes that rocks similar in every way to lavas erupting from modern volcanoes are indeed the products of ancient volcanoes. ... Thus, basically, we take for granted that features in ancient rocks identical to features we can watch forming today were indeed formed by the same process. In short, the origin of ancient rocks can be interpreted in the light of today's processess."

The simple examples chosen by the above authors make this Principle seem trivially self-evident, but we now have counter examples which show it to be, in actual fact, fundamentally invalid. In particular, the gradual transition between the boulder clay and aeolian loess is totally inexplicable in terms of processes now operating, as N.H. Winchell understood very clearly. One cannot even formulate in his imagination a process consistent with the laws of physics which could have given rise to that peculiar situation.

When in the the course of scientific inquiry an assumption proves to be invalid it is normally discarded and alternative hypotheses are considered in its place. But the Uniformity Principle is more than a mere assumption; it is, as we have already seen, an aspiration. It is the prime essential of a rationalistic view of earth's history so to abandon it would be a profound step indeed. Even with the definitive evidence at his own finger tips Winchell could not bring himself to make this clean break. Instead, he remained true to the faith and served later as President of the Geological Society of America.

However the time has come to lay this Principle to rest and admit frankly that mechanisms have operated in the past which were not only unlike those operating today, they even

Opening the Door

defied the laws of physics as we understand them. This is not to say that the mechanisms acted *contrary to nature*; it is only to recognize that our understanding of nature (as contained in the known laws of physics) is somehow incomplete. If this fact were not already clear from Winchell's remarkable report any last lingering doubt must be dispelled as we probe deeper into these deposits and try to gain some insight to the processes that produced them.

We saw earlier that the drift is essentially devoid of fossil residues. Occasionally one finds organic materials gathered into the drift, or in close association with it, but they are deemed to be alien. From what has been learned so far one might also expect the loess to contain very few fossil residues, and at least with respect to vertebrate remains this certainly proves to be the case. Furthermore, again agreeing with expectation, such remains as are found usually occur near the very bottom of the sheet. Evidently the animals were victims of the dust, and they did not long survive after it started to fall. Perhaps we should hear the fruits of firsthand observation on such an important matter, so here is A. L. Lugn discussing this very point [55; p.150]:

" Ancient loess deposits contain few vertebrate fossils, and these are found almost exclusively in the lower few feet of any loess, which was deposited at the very beginning of the age of dust-blowing, or in old loess soils at the top of loesses, which were developed during intervals of non-deposition.... Apparently, the mammals endured the dust-blowing and scarcity of water and forage as long as possible, then many of them died and left their bones entomed in the first (lower) few feet of the dust deposit. Those which surived long enough migrated to more hospitable localities, and ultimately the population may have reached quite distant areas. Later generations returned to repopulate the loess areas when conditions again became favorable, and there was forage and water and soil development."

So the normal location of mammmalian remains is near the bottom of the sheet as one would expect, but the reason why they are sometimes also associated with the colored bands ("old loess soils at the top of loesses") is not apparent. More commonly, only disjointed parts of plant or mammalian residues are ever found at higher levels in the loess, and these only rarely. But in any case, since we are ignorant of the conditions and the circumstances that brought about the deposition and prevailed at the time perhaps we have no firm grounds for anticipating just what might be found within it—or where.

In particular, one would probably not expect to find the shells of snails within the loess, but, as a matter of fact, they abound in it! Indeed, they occur so widely as to constitute a distinctly characteristic feature, and that brings us to one of the most interesting aspects of the entire subject. For while mammalian remains are so rare as to be a genuine curiosity, snail shells occur in great numbers—not everywhere, and not uniformly to be sure, but locally, according to Flint [33; p.253], they have been found in excess of 5,000 to the cubic foot! Moreover, they are to be seen at all levels, from the bottom of the sheet to the very top.

With only rare exceptions these are the shells of land-dwelling snails, and they usually prove to be the same varieties as may be found living around and about the deposit today. Plate 31 displays a few such shells taken from the loess at Council Bluffs. The scale is given by the square frame which is two inches on a side. These particular specimens are not large and neither are they "fossilized"—that is, the original material has not been replaced by silica. In fact, one of their most surprising attributes is that they appear to be absolutely fresh and new. Excepting only for their coloration, no degradation is evident even in the most minute of details.

Certainly these shells pose a very difficult problem for the prevailing theory of loess deposition since snails require vegetation for forage. If the mammals inhabiting the region

Opening the Door

PLATE 31: *Snail shells from the loess near Council Bluffs, Iowa. Approximately twice natural size.*

were all killed off at the beginning of the "age of dust-blowing" then how is it that the snails seem to have survived during the whole time? The general absence of any kind of vegetable remains in the loess becomes an even more pressing problem for the uniformitarian point of view because of this great army of snails. For if the silt accumulated slowly, burying these snails, then why did it not bury the vegetation upon which they fed as well?

Setting this question aside for the moment let us recall those limey nodules widely found in the loess and resolve to look inside some of them. Plate 32 shows sawed sections of two typical samples of different size and composition. Plate 33 shows another—this one with the end broken off to expose the interior, and one can hardly fail to notice a surprising common property. They are not solidly filled, and moreover they tend to have a mud flat texture on the inside. Plate 34 is a closer view of a portion of the previous sample, and it shows this remarkable feature with unambiguous clarity. Even if we cannot say exactly where these nodules came from or the precise

PLATE 32: *Sawed sections of two typical loessian nodules showing interior cracks and cavities.*

« 154 »

Opening the Door

conditions under which, they were formed, nevertheless we can draw some conclusions about their origin from this mud-flat texture. Clearly they were at one time muddy on the inside while they were more nearly dry at the surface. Then, as the mud eventually dried cracks formed inside because the original interior volume could not be preserved in the dry state. Could the nodules have formed in place within the body of the loess?—perhaps at localized voids where water seeping down from the surface might collect to form muddy globules?

One would hardly think so—and for several reasons. Firstly, the loess is too porous to allow water to collect in

PLATE 33: *Loessian nodule with one end broken to expose the mud-flat texture of the interior surface.*

localized globules at any supposed interior voids. Secondly, if there were such interior voids one would not expect them to be globular in shape; they would be more randomly and irregularly shaped. Thirdly, even if the voids were globular in shape, one would expect a nodule formed in this way to reflect the direction of gravity by a gradation in texture from top to bottom—which is clearly absent in these objects. And fourthly, the well-defined surface and the surface texture, especially the wrap-around texture seen in Plate 29, testify that the nodules were already formed as entities before they came to reside in the loess. Evidently, then, they fell fully formed right along

PLATE 34: *Close-up view of the exposed interior surface of the sample of Plate 33.*

Opening the Door

with the silt itself. We had already deduced from the occasional pebbles in the loess, and from those great limey slabs upon the loess in China, and from other reasons as well, that the silt was not normal terrestrial material that had been simply redeposited by the wind. Here is additional testimony to this same effect if more were needed; assuredly these nodules were not blown into place by the wind.

Then the remarkable sample shown in Plate 35 removes all doubt about the origin of the snails; evidently they fell along with the silt as well! Howorth was correct in concluding that

PLATE 35: *Loess nodule with snail shells firmly incorporated upon the outer surface.*

they did not forage upon vegetable matter growing on the silt while it was slowly accumulating. Notice that none of the shells is intact. In particular, little remains of the one to the left of center, but, surprisingly, it was filled with the same material as constitutes the nodule at large.

Plate 36 is a closer view of this interesting specimen. It is worth noting that imprints of the cracks in the now-missing shell can be discerned upon the casting which remains. Evidently the shell broke while the mud was still soft—that is, very shortly after the nodule had formed. Plate 37 gives an even closer look at the imprint of one of these breaks. The white

PLATE 36: *Closer view of one of the shells seen in Plate 35 showing interior filled with nodular material.*

Opening the Door

area at the upper left is, of course, a part of the shell still in place. Notice in this picture that although the mud was still soft enough to be disturbed by the breaking shell, a very thin layer near the surface was quite firm—in fact even brittle since it broke into discrete flakes which did not subsequently deform.

Then why did the shell break at all if it was so well supported from behind? Presumably the break was not a consequence of the impact of the muddy ball with the newly fallen silt for the nodule itself, soft as it was inside, shows no evidence of deformation from the impact, and the firmly supported shell should have been stronger yet. Furthermore, after

PLATE 37: *Close view of the interior casting of a snail shell on a loessian nodule showing imprint of fracture lines.*

the nodule was in place, buried in the silt, the shell would have been completely safe from harm so one must conclude that the shell was filled and the breakage occurred even before it fell into place.

Plate 38 is a closer view of the coiled shell prominent in Plate 35. Careful examination reveals that it has been squeezed, collapsing in the process—note how the successive coils overlap upon one another. But this is only to be expected in light of our present understanding of the origin of these objects. The nodule formed as a ball of mud, and the snail became affixed to the surface at that stage. Later on, the nodule contracted as it

PLATE 38: *A closer view of the coiled specimen visible in Plate 35 showing deformation due to squeezing.*

Opening the Door

lost its moisture, compressing the shell into its present condition. The surface does not show a mud crack texture because no tensions developed there in the drying. Tensions were confined to the interior regions where, as previously concluded, the proportion of water was initially greater.

Plate 39 shows another example of a loess nodule with residues of a snail affixed; the shell in this case is scarcely imbedded in the surface at all, but instead it mounts upon a kind of pedestal composed of the nodular material which extends out and smoothly into the shell. Plate 40 shows this

PLATE 39: *Residues of a snail shell filled with nodular material extending out from the surface of a loess nodule.*

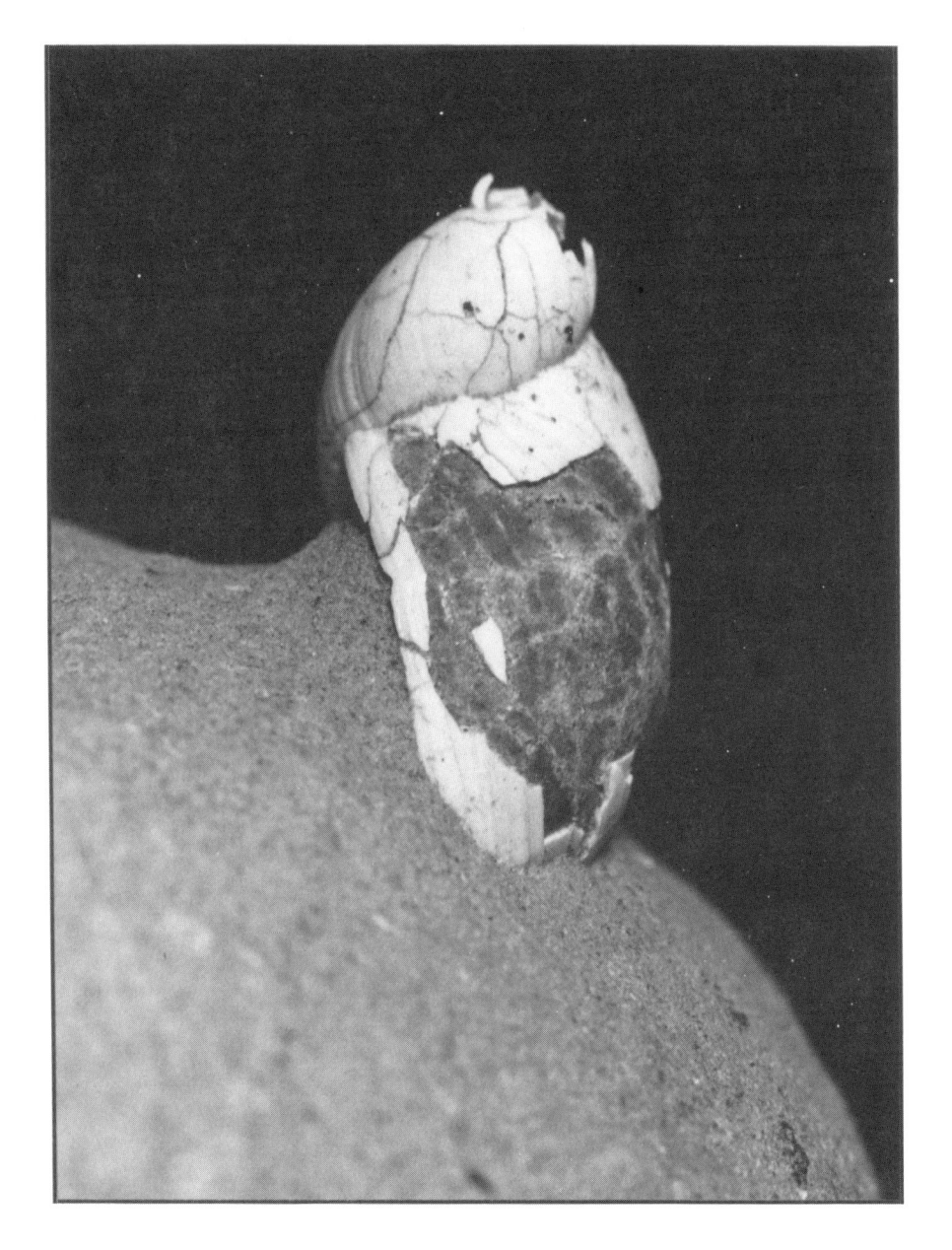

PLATE 40: *A closer view of the specimen in Plate 39 showing the pedestal of nodular material which supports the shell.*

Opening the Door

remarkable feature in greater detail. Note that here again the shell has broken*, and imprints made by the cracked edges of the shell are still to be seen on the surface of the casting. What strange affinity had the mud for the inside of this shell?

From that artificial perspective inside The Laboratory none of these forms make any sense, but stepping outside we gain a broader view which permits, up to a point, an easy resolution to all of these riddles. *We conclude that there exists an added dimension of space† which can come into play abnormally at times. Under such circumstances palpable material can enter into our world along that other dimension, in violation of the customary conservation laws.* Of course, we are not mentally equipped to actually perceive this extension of space or understand how the process works, but our language does have the words to describe it superficially. In this context we say that the alien material enters our world along that other dimension from some "parallel plane of existence". Those who might have witnessed the event, then, watched dumbfounded as the silt and other materials "materialized" out of bare nothingness before their very eyes and dropped gently to the ground.

---

* *The break in the body of the shell is natural as before, but the tip was broken by unfortunate accident while being photographed.*

† *Perhaps it should be emphasized that we are here speaking of an added* space-like *dimension, which is to be distinguished from the commonly recognized* time-like *fourth dimension. Of course the idea of extra space-like dimensions is not new; philosophers, scientists and writers have worried the point for many years. Cosmologists likewise, in their search for a "unified theory of fields", deduce that extra dimensions of space must actually exist. In fact, most recent theory suggests there might be as many as six such extra space-like dimensions. See, for example, Reference 50, passim. These extra dimensions are presently thought to be little more than mathematical curiosities which in no way vitiate the normal laws of physics, but obviously this supposition is not always correct.*

Furthermore, if we go on to assume that net electric charge accompanied the silt then all of the bizarre properties noted in the loess are easily explained—in general terms, if not in detail. For example, as the electrically charged silt particles fell to earth their mutual repulsion would discourage them from settling any closer together than was actually necessary for physical support. Thereby would result the surprisingly porous structure of the undisturbed loess. Then, as the silt fell to earth and began to accumulate, its electric potential would increase, and sparking into the atmosphere would ensue in order to deplete the charge and lower the electric energy. One can identify those ever-present capillary tubules as the paths left by those sparks.

Evidently the loessian nodules developed from globules of water entering our world in this fashion—globules that were disturbed into irregular shapes, possibly by strong turbulence in the surrounding air. During their passage they hovered for a time in an environment where nodular solids were "materializing", and some of this solid substance materialized within the globules themselves, forming mud. One can then picture more of that dust materializing in the surrounding space, and, owing to the turbulence, the dust would be driven into contact with the developing mud, where it would adhere. Thus can one easily visualize how the surface came to acquire a greater proportion of solids than did the interior.

Evidently snails also materialized into this environment, and those that chanced to make contact with the muddy globules at an early stage also adhered. While solids were materializing within the water globules they also formed within the bodies of the snails and turned them into mud as well. The volume of the resulting mud was naturally greater than the volume of the snails themselves so this mud overflowed the shells as we see in Plate 40. Eventually it became too viscous to flow, at which point it simply expanded and the shells cracked—as observed in every case. Apparently a membrane

Opening the Door

just inside the shell prevented the developing mud from adhering to the shell so the broken pieces were free to fall away; presumably vestiges of this membrane gave rise to the superficial flakes visible in Plate 37. And so are all the principal properties of the loess, otherwise hopelessly intractable, easily accounted for within the context of this picture.

William James, we recall, noted a kind of dust cloud of exceptional observations in every science which are easier to ignore than attend to forthrightly. The loessian nodules are certainly one case in point, and in fact their bizarre origin reminds one of a whole class of phenomena popularized by Charles Fort. A writer by profession he took a special interest in those extraordinary events that Science refused to recognize. His plaintively snide critique of a complacent intelligentsia, *The Book of the Damned*, was first published in 1919 [34]. It lists countless noteworthy events—matters of record that had been damned into oblivion simply from having been quietly ignored. But let us here resolve that the time for "sweeping such oddities under the rug" has passed. Instead, let us address them openly and see what can be learned from them.

Most of the events in Fort's compilation involve the fall of unlikely materials from the sky, reminiscent of the fall of loess. In the normal course of events such falls usually occur during severe thunderstorms, and typically they are confined to very small areas. The material involved may be animal, vegetable or mineral, and occasionally substances are seen to fall which do not fit into any of these categories. One such event reported in detail by A. Meek displays most of the typical features; he described it in these words [58]:

" About 3 o'clock on the afternoon of Saturday, August 24 last [1918], the allotment-holders of a small area in Hendon, a southern suburb of Sunderland, were sheltering in their sheds during a heavy thunder-shower, when they observed that small fish were being rained upon the ground. The fish were being precipitated on three adjoining roads and on the allotment-gardens enclosed by the roads; the rain swept them from the roads into the gutters and from the roofs of the sheds into the spouts.

" The phenomenon was recorded in the local newspapers, the fish being described as "sile." I was away at the time, but, seeing the account, I wrote to Dr. Harrison, and thanks to him, and especially to Mr. H. S. Wallace, I obtained a sample of the fish, and I was able yesterday (September 5) to visit the place in the company of the latter gentleman.

" From those who saw the occurrence we derived full information, which left no doubt as to the genuineness of what had been stated, and this we were able to put to the test, for a further sample was obtained from a rain-barrel which could have got its supply only from the spout of the shed to which it was connected. The precipitation of the fish, we were told, lasted about ten minutes, and the area involved Commercial Road, Canon Cocker Street, the portion of Ashley Street lying between these streets, and the adjoining gardens. The area measured approximately 60 yards by 30 yards, and was thus about one-third of an acre. It is not easy to say how many fish fell, but from the accounts it may be gathered they were numerous; there were apparently several hundreds.

" There can be no question, therefore, that at the time stated a large number of small fish were showered over about one-third of an acre during a heavy rain accompanied by thunder; we were informed that no lightning was observed, and that the wind was variable.

" All the examples which came into my hands from different parts of the ground and from the rain-barrel prove to be the lesser sand-eel (*ammodytes tobianus*). They all, moreover, are about 3 in. in length, or 7.5 cm. to 7.9 cm. They are not "Sile", a name usually given to the very small young of the herring. But the sand-eels are sea-fish, and it is evident that the sand-eels showered to ground at Hendon were derived from the sea.

" On sandy beaches around our coasts the lesser sand-eel

Opening the Door

is very common. As its name implies, it burrows into the sand, but in the bays it may often be seen not far from the surface swimming about in immense shoals—shoals which are characterized by the members being all about the same size.

" The place where the sand-eels in question were deposited lies about one-quarter of a mile from the seashore, but it is probable that the minimum distance of transport was at least half a mile.

" The only explanation which appears to satisfy the conditions, therefore, is that a shoal of sand-eels was drawn up by a waterspout which formed in the bay to the southeast of Sunderland, and was carried by an easterly breeze to Hendon, where the fish were released and deposited. It is significant that the area of deposition is so restricted, and that no other area was affected. The origin and the deposition were therefore local.

" We were informed that the fish were all dead, and indeed, stiff and hard, when picked up immediately after the occurence. This serves to detract from the possibility of distribution being influenced by such an occurrence, but it is possible that other species would be able to withstand such an aerial method of dispersion. It is more than probable that the vortical movement of a waterspout would transport plankton. This was naturally not observed in this case, and the small creatures, including eggs and young stages, would likely be carried over a wider area."

The author's closing remarks are especially provocative. If the event had indeed been caused by a waterspout, the only agency which could conceivably fit inside The Laboratory, then other peripheral materials would have been carried along with the fish as well, but nothing of that kind was observed. That the fall of fishes should have continued for ten minutes and yet cover a mere third of an acre also seems especially noteworthy. In light of our new insight one might say that a "hole opened up in space" and allowed the fish to fall through—but why only fish? Whatever the reason, presumably their journey took some little time because they were not only dead but already stiff and hard by the time they arrived in our world. However other occasions have been reported where fishes fell in all stages of freshness—some of them alive and others not only dead but putrid. The following example is not so well reported, but it has a remarkable feature in common with the loess [38]:

" When we first heard the report of a shower of snails having fallen on Thursday week, near Tockington, in this county, we must confess we suspected the tale to be intended as a test of our credulity; but the fact has been subsequently authenticated by so many respectable persons, and having seen from different sources so considerble a number of those little curled light-coloured shells, with a streak of brown, and containing a living fish inside, we feel confident of the truth of the assertion. They fell like a shower of hail, and covered nearly an inch deep, a surface of about three acres, and great numbers were distributed to a much greater extent; shortly after this a storm swept so large a quantity into an adjoining ditch, that they were taken up in shovels-full, and travellers were furnished with what quantity they chose to take, and they were soon carried into the principal towns of this and the surrounding counties!!!"

Here again is a typical Fortean event characterized by a single species of victim—and that in profusion. Whatever the mechanism at work in such falls one is tempted to identify the loessian deposit as the result of several Fortean-like falls running concurrently, each one adding its own narrow range of product to the mixture. Possibly the fact that such a great quantity of material fell had nothing to do with the availability of source material but was instead *a measure of the energy of the stimulus* and the span of time during which it operated. Recalling Keilhack's suggestion that all the world's supply of loess derived from the same "large mixing bowl", one might imagine that the mechanism generating this material was somehow connected to several different "holes in space", each of which delivered forth its product upon a different location.

In Appendix A we deduce that other worlds very likely overlie our own along that other dimension so one is tempted to suppose that they may be involved somehow with these phenomena—a point that we shall pursue in a later context. But for the present let us observe that since Fortean falls tend to be associated with severe thunderstorms perhaps one can deduce that the storms themselves must be of a bizarre nature. Then, since the loessian silt particles materialized carrying an electric charge, it is only a step to recognize that charged water droplets may similarly materialize under suitable conditions. Such electric charge injected into the air would cause the air to expand locally and to rise—even as storm clouds do, in fact, rise and billow into the stratosphere. And, of course, this added electric charge would provide for the lightning as well. Certainly the explanation for atmospheric electricity in vogue today—that storm clouds billow up into the stratosphere owing to thermal energy derived from the sun and that they become electrified by friction between moving masses of air—cannot be credited even for a moment.

Perhaps the easiest way to persuade oneself that normal physics fails in a thunderstorm is to note that energy is not conserved; it is effectively created. Typical thunderstorms can sometimes be observed to form as isolated units, even after sundown, so the energy manifested in the extreme atmospheric turbulence and as lightning has no demonstrable source in this world. Exactly the same considerations must apply to violent windstorms. The sinuous jet stream aloft, the trade winds below, hurricanes and killer tornadoes, having no discernible source of power, must be driven by extraterrestrial influences which somehow manifest in our world. One might suppose the immediate driving forces to be electrical in nature—that ions in the atmosphere are dragged along by electric fields and thereby set the whole mass into motion. In any case one can but feel relieved that no longer must he explain those powerful winds as the cumulative result of gentle updrafts which arise when the earth gives up to the atmosphere the heat that it receives from the sun. In due course we shall gain some insight to the source of that energy, but for the present let us put these questions aside and continue along our way.

Returning, then, to the loess we have seen that it sometimes grades smoothly into boulder clay, but an instance can be cited where it is also merges gradually into sand. Figure 6 illutstrates one such instance in the state of Nebraska, here carefully redrawn from a figure given previously by A. L. Lugn; political boundaries within the state of Nebraska and other details unrelated to either the loess or sand have been omitted for the sake of clarity. Whereas both the loess and till cover vast irregular areas, this sand is well localized in the form of a

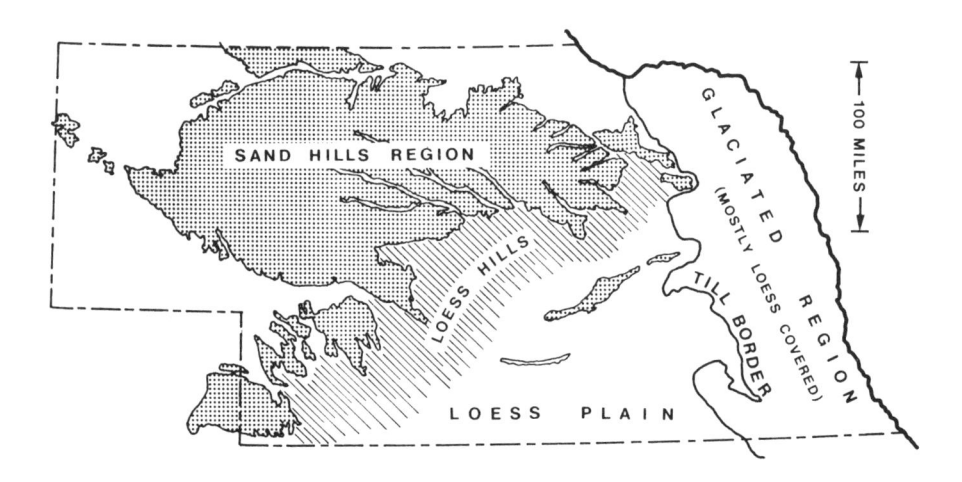

FIGURE 6: *Distribution of sand, loess and drift in Nebraska after A.L. Lugn (Figure 9 - 2 of Reference 55). The sand covers an area in excess of 20,000 square miles, making it the largest sand deposit in the western hemishpere. It is topped with a thin stabilizing layer of soil which has preserved the original teardrop shape intact.*

Opening the Door

well-defined teardrop, suggesting that it was released from a stationary globular region high aloft while the wind blew steadily toward the east and while the loess was also falling nearby; the transition region, shown by the diagonal stripes, is approximately fifty miles wide. The globular body might have been as much as a hundred miles in diameter. Whether this was essentially a Fortean event or a phenomenon more directly related to the drift remains to be seen, but it speaks again of some very intense disruption which overcame the world during that relatively brief period of time. Moreover, it should alert us to the possibility that other great sand deposits might have had a similar origin.

Is it possible to assign a date to this period of great upheaval? One can hardly imagine the loess to be extremely old because both the calcite binding cement which holds the grains together and the snail shells within the mass remain largely intact. During the course of time they must eventually dissolve away as rain water seeps through it, but we have just seen that the finest details of the snail shells show no degradation whatever due to any such loss of calcite. Radiocarbon assays, on the other hand, would indicate a very considerable age indeed if the laboratory results can be interpreted in the customary manner. Ruhe [69], for example, has catalogued many such determinations relating to the loess in Iowa, and a great range of apparent ages is found amongst them. In fact, no radiocarbon whatever could be found in many samples, which would normally indicate an age greater than 47,000 years, the extreme limit of the method. What can be made of such findings?

Evidently radiocarbon analyses cannot be correlated with age in these cases. Very sound reasons compel one to recognize that these events happened quickly, so whenever they occurred the various analyses should yield exactly the same age if the method were working properly. Since the ages do not agree one must conclude that the phenomenon itself undermined the method in some way. Perhaps the unstable radioactive nuclei became even more unstable in that bizarre environment so they decayed away more rapidly than usual during that brief span of time, leaving a depleted (and therefore apparently much older) carbon behind. Of course, radiative decay rates are not affected by conditions that can be achieved in the laboratory, but they have not been tested under these conditions—nor, probably, can they ever be. For the present, then, let us accept the age of the loess as indeterminate and hope that additional insights may gleaned from the discussions which follow.

Chapter 8:

THE GREAT DESTROYER

IT IS NOW PLAIN that the cosmos encompasses at least one added dimension of space which, coming abnormally into play, can greatly disrupt familiar natural behavior. Although we found this simple truth by probing into the loessian nodules it must be a property of nature at large and probably enters to some extent into every natural process whatever. In particular, we deduced that its workings energize thunderheads and drive the winds of passion, and one might anticipate that the basic processes of life itself also depend heavily on this added dimension of reality. Of course, this would require quite a different perception of the phenomenon of life than prevails generally today. Without a doubt this new insight has far-reaching implications indeed, concerning which a few added thoughts are offered in the Epilogue.

Of course, mathematicians can represent spaces in any number of dimensions without difficulty, but even the cleverest of them cannot actually visualize a space of four dimensions. One should not feel uniquely inadequate, therefore, if he also fails in the attempt, but he can construct in his mind a simple analogy as an aid to thinking. Namely, one can picture two-dimensional "worlds" immersed in our three-dimensional space. For example, a tall apartment building might be looked upon as many two-dimensional realms spaced along an added (third) dimension of space, each level being like a world unto itself and going its own way independently of the others. But the various worlds in this building are not truly two-dimensional; they are thin slices in three. And each of them has two neighbors—one above and one below, so it is separated by two boundaries from the adjoining worlds. Perhaps our domain in three dimensions is analogously a narrow slice* in four, and we also may be closed in by two such black boundaries, though we are not able to perceive either of them, and assuredly we cannot see beyond them.

Returning to the task at hand we recall that Ignatius Donnelly traced the drift deposits to a catastrophic collision between the Earth and a comet. He argued (correctly, as we have seen) that the drift deposits could not be explained in terms of any known natural agency. And then he urged that since comets were the last unknown quantity remaining in the arsenal of nature then by elimination the drift must have been caused by a comet. But he also believed that the myths and legends which survive from olden times were not contrived out of whole cloth; even the most fanciful of them, he thought, must have a basis in fact that was understood by the next generations or they simply would not have been propagated. He thought that some of these legends were inspired by cometary visitations, so he tried to formulate a model for comets that would accommodate the drift and the old legends as well. It will be worth our while to review two of the legends that he had in mind.

In particular, he thought that the legend of Phaëton was inspired by such an event. Its origin is lost in antiquity, but we know that Solon, the Athenian law-giver, heard the tale during his travels in Egypt—though apparently it was known in Greece

---

* *This picture is carried a step further in Appendix B where it proves to be especially fruitful.*

The Great Destroyer

even before that time. In essence this is how it goes:

Phaëton was the son of Helios, the god who drives the chariot of the sun across the sky. When the lad came of age he wanted a chance to drive the chariot, but his father would not allow him to do so because a great deal of skill was needed in order to manage the horses properly; they had to be guided carefully or the sun would not follow its proper path across the sky. Moreover altitude had to be maintained just right or the earth could be scorched. But the boy continued his pleadings until at last the father relented. On the selected day Helios gave the boy explicit instructions about all the necessary details and then sent him on his way.

Surely the reader can already guess how the story ends. Phaëton lost control of the horses so the sun dipped too low in the sky, and the earth was burned in a great fire. As the story continues his sisters visited the site where the chariot crashed and cried bitter tears over their brother's body. Legend has it that their tears later turned to amber. Donnelly considered this myth to be a fanciful account of the approach of a blazing comet which burned the earth widely before colliding with it. He surmised that the collision excavated the basins of the Great Lakes, and he identified the drift as debris scattered from the impact.

Another ancient account describing an errant sun is to be found in the Book of Joshua. Chapter 10, starting in verse 12, reads as follows:

" Then spake Joshua to the Lord in the day when the Lord delivered up the Amorites before the children of Israel, and he said in the sight of Israel, Sun, stand thou still upon Gibeon; and thou, Moon, in the valley of Ajalon.
" And the sun stood still and the moon stayed, until the people had avenged themselves upon their enemies. Is this not written in the book of Jasher? So the sun stood still in the midst of heaven and hasted not to go down about a whole day."

Now perhaps one can be forgiven for doubting that the earth actually stopped its rotation for a time and then started spinning again, but some genuine visual effect must have inspired the story or it would not have survived the first reading. This is a significant variation on the previous theme because in standing still for a prolonged period the errant sun could not have been confused with a blazing comet about to strike the earth. Carmack [23;p.129] reports a similar legend among the Quiché Mayans of Central America which tells of a time when the people saw three suns in a single day.

Then, with a view toward explaining the drift deposits, while making sense of these old legends as cometary visitations, Donnelly reasoned backwards to conclude that comets must consist of [30;p.69]:

" First, a more or less solid nucleus, on fire, blazing, glowing.
" Second, vast masses of gas heated to a white heat enveloping the nucleus, and constituting the luminous head, which was in one case fifty times as large as the moon.
" Third, solid materials, constituting the tail (possibly the nucleus also) which are ponderable, which reflect the sun's light, and are carried along under the influence of the nucleus of the comet. ... "

Learned opinion in his day held, simply stated, that "comets are the nearest thing to being nothing at all that anything can be and still be something" so his theory was in trouble from the beginning. But its greatest problem was that it really did not satisfy the need.

In truth, a proper perception of comets proves to be the very key to understanding the face of the earth and its lost history so we must persist. Having seen how Donnelly viewed these objects a hundred years ago, let us now examine the best of current thinking. Suggested by Fred Whipple it is known affectionately as the "dirty snowball" model. According to this view, comets are condensed aggregates of the primordial mat-

The Great Destroyer

ter that originally formed our solar system—namely dust and frozen volatiles of various kinds. The picture, then, is that when one of these objects draws near to the sun the exposed ice evaporates to produce a gentle wind blowing outward from the core, the liberated dust being carried along with the breeze. Ultraviolet radiation from the sun causes the gases to fluoresce and emit a bluish light, usually in the form of a halo around the center. On the other hand the fine dust particles reflect sunlight directly and are driven away from the sun by the force of radiation pressure to form the more impressive dust tail.

More recently, in 1986, a rocket launched space probe, "Giotto", intercepted Halley's comet and returned copious physical measurements of the evolved gases and fairly good photographs of the nucleus [14]. These showed a vague, stark black "peanut shaped" body measuring some eleven miles in the long direction and enclosing a volume of about 26 cubic miles. The sooty blackness of the nucleus came as a surprise and helps to explain why no sign of a nucleus is normally seen, even with large telescopes; the albedo was only about 4%, making it the blackest celestial object ever observed. But most surprising of all was the fact that its *dust and gases were emitted from a few well-defined jets* while some 90% of the surface was entirely inactive. Clearly, the dirty snowball picture has no place for these improbable jets—which gives our first hint that the true nature of comets is yet to be discovered.

Continuing with our summary of prevailing views, the theory of their origin proposed by J.H. Oort (see Reference 64, for example) enjoys wide popular support. He suggested that vast numbers of these condensations of primordial matter, left over from the formation of the solar system, still continue to circulate in regions so remote from the sun as to be within interactive range of the nearby stars. Because of these interactions their angular momenta about the sun would have become randomly distributed, both in magnitude and direction. According to this picture those that happen to acquire close to zero angular momentum about the sun as a result of these stellar interactions must begin to fall toward the sun and will eventually show up as comets.

We shall find it instructive to test this hypothesis for ourselves and determine if the newly arrived comets from this proposed reservoir in space are indeed distributed randomly in regard to their angular momentum about the sun as Oort's model requires. In principle, the members of this "cloud" of new comets should be distinguishable by their extremely long period of revolution, a property associated with their supposed origin in remote space. As comets pass through the planetary system they exchange energy with the planets and thereby settle into shorter period orbits where they become more decidedly members of the sun's family. The orbital properties of these short-period comets are therefore to a greater extent *acquired* properties so they would not be of interest for this survey. For this test, then, let us agree to consider only those comets with periods in excess of 100 years.

But the experiment has hardly begun when we encounter a slight complication. Since comets are less easily distinguishable at greater distances from the sun, those with large perihelion distances are more likely to pass unnoticed and thereby miss being represented in our survey. Since the distance of closest approach to the sun is actually a measure of angular momentum the inherent observational bias in favor of small perihelion distance is likewise a bias in favor of small angular momentum as well. In an effort to limit the effects of this bias in our study let us further agree to consider only those comets that were observed during the present century. During these more recent times of intensive celestial observations comets are routinely discovered at greater distances from the sun so our sampling should be valid over a larger range of perihelion distances. Of course, observational bias at very large perihelion distances would become increasingly severe in any case. The Catalogue of Cometary Orbits [56] lists 235 long-

The Great Destroyer

period comets which passed perihelion since January of 1901.

Proceeding with the analysis let us construct in our imagination an artificial space—an "angular momentum space" in three dimensions in which the angular momentum about the sun of each of these comets is plotted as a single point (since this quantity is a constant of the motion). And then suppose that we divide up this space into a series of concentric shells, all of the same thickness, centered on the zero point. Then the volume of any given shell will be proportional to the square of its average radius, and if the points representing comets are randomly distributed throughout this space as Oort's model requires, then on average the number of points to be found in any such shell will also be in proportion to the square of the radius.

Of course radius in this context corresponds to magnitude of angular momentum, so in plain words: Observational bias aside, according to Oort's model which requires a random distribution in angular momentum for primeval comets, the number of comets observed in any constant small interval in angular momentum should be in proportion to the square of the angular momentum itself. The histogram in Figure 7 shows the distribution in angular momentum of these 235 comets. The units are arbitrary; angular momentum is calculated here simply as the square root of the perihelion distance as expressed in astronomical units.

If one assumes no observational bias for perihelion distances less than unity then the first ten counts on this plot effectively define the average density of points in angular momentum space, and this should allow us to extrapolate to larger radii and calculate the number of comets that should have been observed in the various intervals having angular momentum greater than unity. The parabolic curve in Figure 7 increases as the square of the angular momentum as Oort's model requires and is a plausible fit to these first ten points. It therefore describes how the comet count should have varied

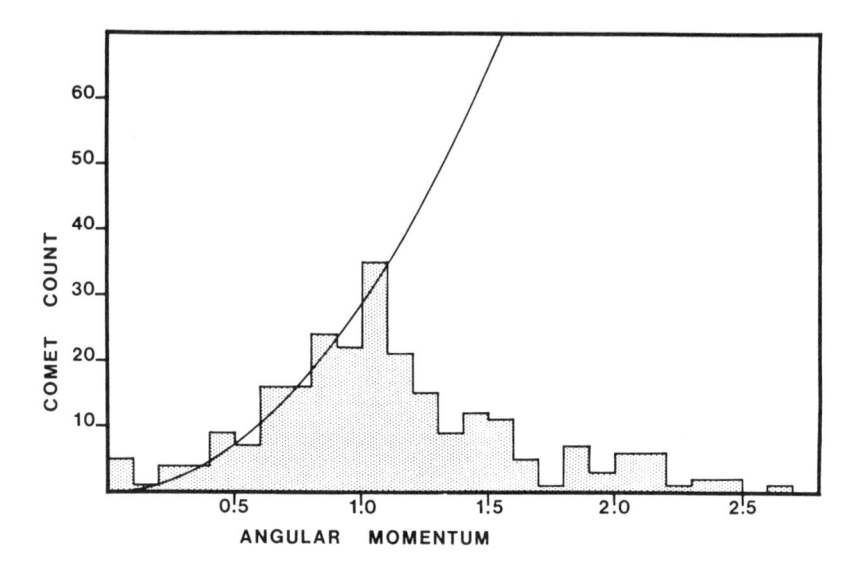

FIGURE 7: *Histogram showing the distribution in angular momentum of the 235 long-period comets observed since January of 1901. Angular momentum is here expressed simply as the square root of the perihelion distance in astronomical units. The parabolic curve shows how the comet count should have increased at values greater than 1.0 if their angular momenta were actually distributed randomly as required by Oort's model.*

with increasing angular momentum—apart from observational bias of course.

Quite evidently the observed counts in the region greater than unity bear no similarity whatever to what would be expected from Oort's model, and the disagreement is much too profound to be attributed to mere observational bias. Probably everyone who has ever undertaken a serious study of comets has made a plot similar to Figure 7 for himself, and having done so has quietly set it aside and tried to forget—perhaps because its implications conflict so sharply with the prevailing theory of origins. For the fact that comets as a class possess such small

The Great Destroyer

angular momentum about the sun implies that *they must have been inherently associated with it from their birth, and that can only mean that they originated at the sun itself.* In that case, in their initial states they possessed no angular momentum at all; whatever angular momentum they now possess was gained from gravitational interactions with the planets.

This may seem only an imperceptible improvement over the former model. Whereas Oort pictured aggregations left over from the beginning of the solar system we now have to think of materials thrown out from the sun—possibly during some uncommonly powerful solar flare. One might suppose that the hot gasses condensed, and then in the course of time weak gravitational forces drew the debris together, giving rise to a dirty snowball as before. But this picture discards too quickly the fact that, originating from the sun, those ejecta possessed enormous self energy. Consequently, we have only to modify the picture slightly in order to define a comet that will do all that Donnelly required.

Reasoning backwards then, from the observed effect to deduce a cause, we have to suppose that those ejecta occupy a special metastable state—*one that possesses no mechanism for the loss of thermal energy to the surrounding space.* The nature of this hypothetical state is by no means obvious, and in a cosmos of three dimensions would be entirely unthinkable. But we now understand that there exists at least one added dimension of space so what would otherwise be an insurmountable hurdle reduces to a mere conceptual difficulty.

Next, in order to agree with the normal appearance of comets one must conclude that this special state has a kind of "axial" structure and that material descending from the metastable state into our world-plane passes along this axis and becomes visible only in the restricted region where the said axis intersects our plane of existence—regardless of the actual size of the comet itself, which could be huge. This region is here called the active eye. Since no energy is lost by radiation the stored material issues from this active eye at an unpredictable rate but always at a temperature close to 6000° K, namely the surface temperature of the sun.

And then, in order to accommodate that hovering behavior seen at the time of Joshua one has to conclude that upon striking a planet a comet retains its integrity and may become affixed in some way to the surface. If one were reasoning in the other direction, from cause to effect, he might anticipate that, occupying another dimension of space the object would not interact with the planet at all but would pass through unaffected and continue on its way. However, in order to accord with those ancient accounts he has to conclude that there is some interaction after all—an interaction that brings the object to rest near the surface, though probably extending somewhat beneath it. Since they survive the impact intact presumably such collisions are relatively tranquil. Possibly the objects manifest very little mass in that metastable state (and therefore possess little kinetic energy) which is how available forces on the sun were able to accelerate them to near escape velocity at their birth.

However that may be, evidently their gravitational mass and inertial mass remain exactly equal so that they move under the influence of gravity precisely as do normal material bodies. In that case, after being ejected from the sun a comet might mature somewhat in this fashion: As hot gases (actually a plasma) emerge from the active eye solid condensate accumulates in the vicinity and eventually coalesces to form a ponderable nucleus surrounding the eye. As this nucleus grows in size its periphery must cool—ultimately to the point where even some of the more volatile components condense, binding the assemblage together all the more firmly. Thus, as the nucleus continues to grow it acquires the character of Whipple's dirty snowball, but we have to keep in mind that the snowball is to be distinguished from the comet itself. Nevertheless it is interesting to note that this snowball would display some cometary

The Great Destroyer

properties, in keeping with Whipple's picture, even after the comet itself had faded away.

The most volatile of the gases would continue to escape into outer space, but as the interstices within the collecting mass became plugged with ice the number of open channels would continually decrease—to the point where only a few would remain, at which time the gases would emerge as discrete jets. Here, then, is an easy explanation for the jets observed on the nucleus of Halley's Comet. Furthermore, we might note that as the channels became more and more restricted owing to the accumulating ice the internal pressure would increase and eventually force the weakly bound structure apart, thereby opening new channels so materials would emerge through an entirely different pattern of jets. Such changes in cometary aspect have been routinely observed, of course, and account for much of the mystique that comets have acquired through the years.

Now according to this picture the plasma emerges from the active eye at approximately the surface temperature of the sun (about 6000°K), but since it must cool on coming into contact with the growing nucleus that original high temperature cannot be observed directly. Nevertheless, Wurm found remarkable indirect confirmation of a high source temperature in the spectra of some comets. He wrote as follows [84; p. 583]:

> "... A completely symmetrical molecule such as C_2 has no permanent dipole, and, to a high degree of approximation, it is unable to emit vibrational or rotational energy in the electronic ground state (or quasi-ground states). The CN molecules, in contrast to C_2, can emit vibrational and rotational energy and, if weakly excited, tend to accumulate in low quantum states, whereas C_2 molecules do not. The CN molecules, for this reason, will simulate a very low-temperature intensity distribution. The C_2 molecules, on the other hand, will keep, more or less, their initial distribution. For all unsymmetrical molecules, we observe only those lines [in the spectrum] with relatively low rotational

quantum numbers and a vibrational distribution that is completely consistent with the proposition that excitation occurs only for molecules in the lowest vibration level. According to Wurm, from a comparison with the distribution in the electric furnace at 2000°K — 2500°K the C_2 molecule shows a vibrational and rotation distribution which corresponds to a temperature of about 2500°K. The origin of this internal energy we do not yet know."

Since this observation is so profoundly important perhaps a few added words of explanation would be of value. Firstly, one fruit of quantum mechanics is that nature does not allow a molecule to rotate or vibrate at just any speed or amplitude. Effectively only particular speeds and amplitudes are allowed; these constitute the allowed energy states of rotation and vibration respectively. Secondly, a basic fruit of statistical mechanics is that such rotators and vibrators will be excited into the higher energy states more readily the higher the temperature. At low temperatures an assemblage of such molecules would all be found in the lowest energy state, but as the temperature is increased more and more would be found in the higher states. In thermal equilibrium the distribution of molecules in the various states is strictly defined by the temperature and the laws of statistical mechanics.

Now the light emitted by C_2 molecules in a comet is actually a fluorescence radiation, stimulated by illumination from the sun. This emission derives from transitions between the various electronic energy states of the molecules, but the energies of these electronic states depend, to a slight extent, upon the states of rotation and vibration of the molecule as well. Therefore, by observing in the spectrum the distribution into the various rotation and vibration states one can deduce the source temperature.

As Wurm pointed out C_2 molecules are symmetrical so they have no dipolar properties in the electronic ground state, and therefore they cannot radiate away their energy of rotation

The Great Destroyer

or vibration. Apparently they are also slow to transmit this portion of their energy by collision with their neighbors or other surrounding materials, for even after they have worked their way through the accumulated nucleus the C_2 molecules still retain much of the thermal energy that resided in these states initially. Therefore Wurm's observation provides unequivocal proof that the effusion from comets first appears at an exceedingly high temperature, and it thereby offers gratifying corroboration for our new model of comets and their origin.

It is also interesting to note that this picture gives an easy accounting for the origin of the earth's magnetic field as a residue of cometary impact. The present point of view which ascribes this field to a self-excited dynamo operating within the earth has always posed insuperable problems, the most troubling of which is that the terrestrial dipole is not symmetrically located. Not only does its axis not coincide with the axis of rotation, but it is also displaced from the center of the earth by several hundreds of miles. Moreover, from an analysis of the residual magnetization of ancient lavas and rocks one can deduce that on several occasions it has unaccountably reversed its polarity. These riddles can now be easily resolved, but as this subject is peripheral to our central theme, and requires more than the usual care in reading, it has been set aside as a topic of its own in Appendix C.

In summary then, taking the event that gave rise to the drift as a model, one must conclude that a collision between a comet and a planet is a multifaceted phenomenon. In the first place, the ponderable nucleus would interact simply as a massive meteorite—explosively. Donnelly may well have been correct when he identified the basins of the Great Lakes as impact craters gouged out by such a collision. And then secondly, apparently the comet itself, manifesting little mass, comes to rest more quietly, but in decaying the active eye spews out incandescent gases from which various solid materials condense. In Appendix D we examine the possibility that volcanism and seismic activity result from captured comets whose eyes happen to reside below ground level. If the eye remains above ground then the condensate would fall to earth as a superficial deposit; presumably the boulder clay resulted from this phase of such an event. And thirdly there may be various peripheral falls of a Fortean nature, arising not from the comet itself, but generated in some fashion by the resulting wrenching of space. Presumably the fall of loess was a peripheral fall of this kind—as was the fall of amber after the event that inspired the legend of Phaëton. Finally, however, let it be noted that the factors which govern the *rate of decay* of these objects remain entirely unspecified by the model.

We might review Figure 6 on Page 170 in this new light. There we saw that the loess grades uniformly into typical dune sand over a space of about 50 miles. Evidently the sand also fell at the same time as the loess. Perhaps it was a direct product of the comet itself, but more likely, perhaps, it was another peripheral Fortean effect that resulted from the twisting of space. Obviously if such a thing happened once then it could happen again so we gain here an easy accounting for the many accumulations of dune sand found throughout the world—and other similarly inexplicable accumulations as well. Great hills of gravel, for example, presently thought to have derived in some undefined way from glacial action, can now be traced to this same mechanism.

Another interesting corroboration of these findings can be gleaned from ancient records that were recently exhumed from the archives by Immanuel Velikovsky. His book, *Worlds in Collision* [81], raised a storm of controversy because he proposed a catastrophic view of history that clashed with many popular ideas. He also identified the great destroyer as a comet, but many of his conclusions proved untenable since, even as Donnelly and Beaumont [8] before him, he erred seriously in his perception of those heavenly wonders. Nevertheless, his most interesting discovery should not be ignored on that ac-

The Great Destroyer

count; it needs only to be properly interpreted. In particular, he found that in ancient times the planet Venus wore a beard and gave every outward sign of being a comet.

He went on to conclude that sometime during the first half of the second millenium B.C. a great comet passed close to the earth and was perturbed into a much closer orbit about the sun. He conceived that this close encounter wreaked havoc on the earth and inspired the myth of Phaëton, already noted. He then supposed that it settled into its present nearly circlular orbit through gravitational interactions with the other planets after threatening the earth several more times. Finally, then, as its cometary features gradually faded, it became the planet Venus that we know today. He defended this idea by citing ancient clay tablets, excavated from the ruins of Nineveh, upon which were inscribed detailed observations of Venus extending over several centuries. These showed an erratic motion not at all consistent with the simple orbit of that planet today.

One obvious reason why this picture was uniformly discounted by astronomers is that the mass of Venus is totally inconsistent with the masses observed for comets. In fact, its mass is only slightly less than that of the earth, whereas no mass whatever has ever been discerned in a comet. And one might well wonder whether so massive a body as Venus could exhibit cometary features in any case. We know that normally the dust and gas evolved by a comet are driven back, away from the sun, by the minute pressure of solar irradiation on the one hand and the solar wind on the other. But both the dust and the gas would be gravitationally bound to a body of substantial mass, so whatever its nature or structure such a body presumably could not exhibit cometary features. However, setting this problem aside for the present let us proceed to examine some of the reasons why Velikovsky thought that Venus wore a beard in ancient times.

Trying to follow his argument can be frustrating because one must endure endless disjointed citations. In fact, he cites most of his sources so briefly as to render them essentially valueless individually. He strives to be persuasive instead by the sheer volume of such citatations. And he is indeed persuasive, but to review his presentation here would be impossibly tedious. Let us be content, therefore, to examine two of his sources which carry some weight by themselves. One of these is Humboldt's report on the legends of the Indians in Mexico whom he interviewed in about 1800 during his travels abroad. Velikovsky reproduced the key passage here as follows [p.163]:

" The star that smoked, *la estrella que humeava*, was *Sitlae choloha*, which the Spaniards call Venus.
" Now I ask, what optical illusion could give Venus the appearance of a star throwing out smoke?"

The meaning of this citation is hardly open to question, and his other good source is in the Book of Job, Chapter 38 and verse 32. The King James Version renders it:

" Canst thou bring forth Mazzaroth in his season? Canst thou guide arcturus with his sons?"

Obviously this means nothing whatever, and other, newer translations from the Hebrew are no better. But the Septuagint render it much differently. Of course, this is a translation of the Hebrew Scriptures into the Greek language executed by seventy Jewish scholars in the third century B.C. Tanslating the corresponding Greek text into English gives [p.202]:

" Canst thou bring forth Mazzaroth in his season and guide the Evening Star by his long hair?"

the meaning of which is also unmistakable.

A suggestion of the bearded Venus was also observed by Plutarch in Egypt. He recorded an inscription that he saw on the front of the temple of Isis at Saïs, and he described this goddess as the equivalent of Minerva. His original Greek was inscribed in the form of a flaring triangle, and Higgins pub-

The Great Destroyer

lished an English version in this same form which reads as follows [46;p.311]:

> I Isis am all that has
> been, that is or shall
> be; no mortal man
> hath ever
> me un-
> vei-
> le-
> d

Except for the name itself this might easily refer to Venus with a beard; indeed, one would be hard pressed to interpret it in any other way. Even the unexpected shape of the inscription is suggestive—a well-defined point at the bottom with a "beard" billowing upwards. Let us keep in mind that Venus is only visible when the sun is below the horizon so its cometary tail, which would always point away from the sun, would extend generally upwards.

But the difference in name is not at all significant since the gods were known by different names in different countries. And perhaps we can deduce how this variation came about. Namely, according to Plutarch Isis was the equivalent of the Roman Minerva who, in turn, is recognized as the equal of the Greek Athena. But Athena, according to ancient legend, is said to have sprung full grown from the brow of Zeus. Many generations of scholars have tried to find some plausible basis for this legend, but an easy interpretation comes readily to mind in our context; we simply conclude that Zeus was a comet that was seen to divide, a common enough phenomenon, and the daughter comet was dubbed Athena. One would have to be forcefully restrained from generalizing on this clue and supposing that common ancient custom was to look upon new and fearsome comets as gods and to give them names—presumably names that described some fancied attribute of the celestial object. In that case we can understand how a particular comet/

god would acquire a different name in each country.

So it seems evident that Venus did indeed wear a beard in ancient times, but in our present light this does not require that the planet descended from a comet; far more likely it is that a comet struck the planet and adhered. But even granting this much it remains to be explained how that massive combination could display cometary properties—properties which could presumably be seen only in gravity-free space.

Taking a hint from the loess one is tempted to conclude that electrical forces lay at the heart of the phenomenon. Namely, we saw that the loessian silt particles carried an electrical charge when they emerged into our world space, and presumably the underlying physics would apply quite generally even if we do not understand it. We also saw how electrically charged water droplets similarly emerge into our atmosphere, thereby energizing thunderheads. Because of this electrical charge, and the resulting expansion of the air, the clouds are driven upwards by the force of buoyancy until the condensation of additional droplets raises the density and brings the upward motion to a stop. One can easily imagine similar processes at work on Venus after the comet collided. Charged dusty clouds would rise like thunderheads—except that in the absence of condensation they would continue to rise so long as turbulence in the air kept the dust in suspension. Indeed, one might picture this as a continuing process with dust-laden air rising in columns like smoke-laden air in a tall chimney. Eventually part of that charge would be repelled into outer space, taking dust with it—where it would be picked up by the solar wind and carried away.

Now in order to understand the erratic motion of Venus as described on the ancient clay tablets we have only to recognize that there prevailed at the time *two* celestial bodies of interest, namely a comet/god called Venus that had been perturbed into an orbit threatening the earth, and a planet, the second from the sun, whose name we do not know. The ancient

The Great Destroyer

astronomers, desperately concerned that the comet might strike the earth, kept close watch in an attempt to anticipate its future course. However, as it happened, it struck the second planet instead. At that point there remained only one celestial body of interest but still two names. Which of them would be kept? It seems no less than obvious that since that single body preserved the appearance of a comet the comet's name would prevail. So the planet came to be called Venus, but of course it was not the planet that had moved erratically. Judging from Velikovsky's findings [p.203] that collision took place in the vicinity of 700 B.C.

One can easily understand how the sight of Venus with a beard might have influenced various religions throughout the world. Since the planet rotates slowly on its axis the beard would be periodically oriented differently with respect to the sun so the configuration of the beard would change periodically as well. This change, coupled with the changing position of Venus with respect to the earth, would give rise to widely changing appearances. At times the beard would divide; part of it would stream out from one side of the planet and part from the other. On such occasions it would appear as horns pointing up and away from the sun, thereby presenting to observers on earth an aspect not unlike the head of a bull.

Velikovsky cited a number of ancient references to a horned Venus, and he even noted Babylonian astronomical records that followed the changing appearance of the horns with respect to each other [81;p.166]. It is interesting that later figures of Isis* always show her wearing a headdress in the

---

* *According to the prevailing calendar of ancient Egypt the Exodus out of Egypt occurred during the reign of Ramses II, the second king of the 19th Dynasty, whereas figures of Isis wearing the globe-and-horns headdress are routinely found from the 18th Dynasty. If the Egyptian calendar is correct then this view would require that the comet had already struck Venus even before the Exodus.*

shape of a globe with horns. This illusion may well have been the basis for the cult of bull worship which was so widespread throughout the ancient world.

Although the beard is gone, some residue of the phenomenon seems to have survived until quite recently and may even be seen today occasionally as the so-called "ashen light" of Venus. Flammarian [32;p.276] describes it as a "mauvish glow sometimes seen inside the crescent". One might liken it to moonglow, but of course Venus has no moon. Since the glow is only rarely visible and has no obvious explanation, many authorities deny it altogether. Flammarian, for example, says that it is seen only through refracting telescopes and stems from flaws in the objective lens. However, Cruikshank [25] joins others who have observed it and pronounces it genuine. Most plausibly it is the last faint gasp of the comet that struck Venus nearly three thousand years ago. Recent measurements by space probes to Venus give further evidence that just such an event did take place; they show exactly what one would expect after several thousand years with an on-board "heater" spewing incandescent dust and exotic gases into its atmosphere. The planet is totally dead; the surface is largely covered with wind-blown sand; the atmosphere is laden with dust, and its temperature approaches 900 degrees Fahrenheit!

However Velikovsky (80;passim) has argued persuasively from detailed mural carvings in the temple at Karnak that the prevailing Egyptian chronology is grossly in error and that, in fact, Queen Hatshepsut of the 18th Dynasty must be identified as the Biblical Queen of Sheba who paid a state visit to King Solomon some 500 years after the Exodus! In that case the most plausible setting for the Exodus is near the end of the 13th Dynasty, which marked the end of the Middle Kingdom in Egypt. Given this reconstruction for the Egyptian chronology, the scenario offered here is perfectly consistent with the use of the globe-and-horns headdress during the 18th Dynasty.

Chapter 9:

A Fateful Rendezvous

WE RECALL THAT WHEN Father Kino first arrived at the ruins of Totonteac (Casa Grande) the Indians then living around and about were unable to explain who those former inhabitants were or what had become of them. Indeed, they may have known nothing whatever, but it also seems possible that communication between the natives and the erudite Austrian simply broke down when challenged with such an unusual event. And even if the Indians did accurately describe what happened Kino might easily have dismissed the account as mere superstition, not even according it a place in his journal. Such a refusal to accredit what is beyond the mundane would be typical of many even today. Indeed, this might be deemed the very hallmark of the scientific man—that which sets him apart from the untutored masses. But erudition is not without its pitfalls, as experience shows time and time again.

Another case in point is offered by the Great Chicago fire of 1871. History records that disaster as having started when Mrs. O'Leary's cow kicked over a lantern in the stable, but here again erudite historians have trivialized the non-trival. In his book, *Ragnarök: the Age of Fire and Gravel* [30], Ignatius Donnelly reported many totally inexplicable circumstances surrounding that fire, not the least of which was that (p.413)

" ... at half past nine o'clock in the evening, *at apparently the same moment,* at points hundreds of miles apart, in three different States, Wisconsin, Michigan and Illinois, fires of a most peculiar and devastating kind broke out, so far as we know, by spontaneous combustion."

A few of those "most peculiar" aspects will be recounted presently because if Donnelly was correct in his deductions those fires and the bizarre circumstances associated with them were caused by a piece of Biela's comet as it impacted upon the earth. This comet was discovered in 1826 and upon subsequent observation proved to have a period of six and three-quarters years, and, ominously, its orbit passed within 20,000 miles of that of the earth. Its next three returns were normal except that in 1846 it passed so near to the earth that it split into at least two parts, and then it disappeared entirely. There was no trace of it during its next three returns, but of course, as a result of that interaction with the earth, both its period and its orbit would have changed somewhat—and slightly differently for each piece. One such piece was observed to brush the earth in November of 1872, but it is certainly possible that another, having been perturbed into a somewhat different orbit, could have collided in October of the previous year.

In that case it will be of interest to note some of the peculiar features of those fires. Here is one of several descriptions reproduced from Donnelly, (emphasis and all) who, in turn, credits Reference [72] as his source [30;p.415]:

" Much has been said of the intense heat of the fires which destroyed Peshtigo, Menekaune, Williamsville, etc., but all that has been said can give the stranger but a faint conception of the reality. The heat has been compared to that engendered by a flame concentrated on an object by a blow-pipe; but even that would not account for some of the phenomena. For instance we have in our possession a copper cent taken from the pocket of a dead man in the Peshtigo Sugar Bush, which will illustrate our

A Fateful Rendezvous

point. *This cent has been partially fused,* but still retains its round form, and the inscription upon it is legible. Others in the same pocket were partially *melted,* and yet *the clothing and the body of the man were not even singed.* We do not know in what way to account for this, unless, as is asserted by some, the tornado and fire were accompanied by electrical phenomena. ...
"   It is the universal testimony that the prevailing idea among the people was, that the last day had come. Accustomed as they were to fire, nothing like this had ever been known. They could give no other interpretation to this ominous roar, this *bursting of the sky with flame, and this dropping down of fire out of the very heavens,* consuming instantly everything it touched.
"   No two gave a like description of the great tornado as it smote and devoured the village. It seemed as if 'the fiery fiends of hell had been loosened,' says one. 'It came in great sheeted *flames from heaven.*' says another. 'There was *a pitiless rain of fire and* SAND.' 'The atmosphere was all afire.' Some speak of *'great balls of fire unrolling and shooting forth in streams.'* The fire leaped over roofs and trees and ignited whole streets at once. No one could stand before the blast. It was a race with death ... ."

According to the picture of comets derived in the last chapter gases and plasma issue forth from comets at a temperature approaching 6000°K, that prevailing at the surface of the sun. So as incredibly hot as were those "flames from heaven", the present model easily accommodates them. We also concluded that the plasma issuing from comets must have a strong magnetic structure so one can understand the origin of those electrical phenomena that fused the coins in the dead man's pockets. Likewise the falling sand comes as no surprise. Notice that in Peshtigo tornado-like winds accompanied the phenomenon, and in Chicago the effect was much the same. Donnelly continues.
"   The huge stone and brick structures melted before the fierceness of the flames as a snowflake melts and disappears in

water, and almost as quickly. Six story buildings would take fire and *disappear for ever from sight in five minutes by the watch.* ... The fire also doubled on its track at the great Union Depot and burned half a mile southward *in the very teeth of the gale*—a gale which blew a perfect tornado, and in which no vessel could have lived on the lake. ... *Strange, fantastiac fires of blue, red and green played along the cornices of buildings."*

Donnelly then goes on to cite still another witness to this torando-like fury and the great heat of the blast (p.422):

" The fire was accompanied by the fiercest tornado of wind ever known to blow here. ...

" The most striking peculiarity of the fire was its intense heat. Nothing exposed to it escaped. Amid the hundreds of acres left bare there is not to be found a piece of wood of any description, and, *unlike most fires, it left nothing half burned.* ... The fire swept the streets of all the ordinary dust and rubbish, consuming it instantly....

" The intensity of the heat may be judged, and the thorough combustion of everything wooden may be understood, when we state that in the yard of one of the large agricultural-implement factories was stacked some hundreds of tons of pig-iron. This iron was two hundred feet from any building. To the south of it was the river, one hundred and fifty feet wide. No large building but the factory was in the immediate vicinity of the fire. Yet, so great was the heat, that *this pile of iron melted and ran, and is now in one large and nearly solid mass."*

Having this picture in mind we might recall how Juan Mateo Manje described the Casa Grande during his visit in 1697 (p. 54):

" ... with walls ... so smooth inside that they looked like brushed wood and so polished that they shone like Puebla earthenware. ..."

But Puebla earthenware *was fired*, and, as is now plain, so also was that great house, though probably for only a few seconds.

A Fateful Rendezvous

The resulting glaze was therefore so thin that it did not survive the centuries. Also, we need no longer blame vandals in the neighboring tribes for having set fire to the roof of that house.

Now then, with this new insight as a guide, let us return to Papago Park and examine a few more residues of the catastrophe that befell those ancient cities. We shall not probe its depths, to be sure, but perhaps we can gain a modest feeling for the phenomenon and distinguish some of its properties.

We recall that the rocky buttes in Papago Park displayed clear flow patterns, suggesting that they had been superficially liquified at one time. Given what we now know about comets a

PLATE 41: *The head region of Camelback Mountain as seen from the south showing cascade features frozen in place.*

process of this kind could easily have manifested a high enough temperature to melt the rocks, but this does not appear to have been melting in the usual sense. Plate 41 illustrates the conceptual problem involved. Here we see the head region of Camelback Mountain. Evidently a large section of it flowed briefly at one time, and then it suddenly returned to the solid condition; note that the cascasde feature midway up the slope is frozen in place. If that great mass of rock had been heated to the melting point one would probably not have expected it to cool so quickly as to preserve this fine detail.

And this was not an isolated case for one can observe that

PLATE 42: *A site on McDowell Butte showing liquid rock frozen into place as it was flowing out of the cave.*

A Fateful Rendezvous

same unlikely behavior repeatedly in the most surprising circumstances. In some cases it least, it appears that the caves, which are so characteristic of the buttes in Papago Park, resulted when discrete pockets of rock turned to liquid and flowed away! Plate 42 shows an example. This site has been seen before from a different viewpoint; it is the second cave up from the "curtained" cave seen in Plate 20. At the left in Plate 42 one can make out the residues of that sheet of melt-rock running down the back wall of the cave, but the structure on the right, which appears to be liquid flowing out of the hole, is something else altogether. The tongue-like extension from the

PLATE 43: *A prominent cave in the northwestern region of McDowell Butte which displays the tongue-in-mouth feature.*

bottom of the cave continues smoothly back into the cave itself; it has not been "added on". This material also appears to have flowed freely, and it too stopped suddenly in flight! In fact, similar features have been met before.

Plate 43 is another view of a region shown earlier in Plate 21. Here again the melt seems to be flowing out of the cave, and the combination presents a "tongue-in-mouth" appearance that can be identified at quite a number of sites around the park. Perhaps it should be emphasized that these tongue-like appendages have typically the same structure and rigidity as does the rest of the surface of the hill.

PLATE 44: *Caves in McDowell Butte showing especially remarkable tongue-in-mouth features.*

A Fateful Rendezvous

Plate 44 shows an especially remarkable example of this tongue-in-mouth structure; these caves are situated in the small cleft in the southwestern area of McDowell Butte. The one near the upper center appears to be drooling badly, and the tongue-in-mouth appearance of the one below and to the left is very clear. However it is interesting that the tongue hangs out to the side at a sharp angle and also that the floor of the cave is correspondingly tilted. Moreover, the drool on the right and the tongue on the left actually intersect, but notice carefully that these two flows did not combine to form a single stream. Somehow they remained separate, for one can clearly make out an isolated but distinct element of the drool well down toward the tip of the tongue.

The enlarged perspective of nature gained from our study of the loess permits an interpretation of these strange features and suggests a possible accounting for this melting phenomenon at the same time. Namely, it seems clear that the two fluid streams discernible in Plate 44 flowed independently even though they occupied the same morsel of space as we know it. This suggests that the rocks flowed when they were displaced somewhat—by whatever means—from our worldly space along this other dimension. Then the flow ceased when the material came back "home". Recalling that our world may be closed in by two such black boundaries, these masses may have been displaced in opposite directions along that fourth dimension. This would be one way that they could have flowed without merging into a single stream; they occupied different spaces altogether while in the fluid state.

As already noted, this tongue-in-mouth feature can be seen in quite a number of caves around the park, but it is by no means common to them all. In light of these considerations perhaps one might regard the "tongueless" caves as the holes that remained when localized portions of the rock were transported along this other dimension and did not return.

Plate 45 dramatically illustrates the complexity of the

PLATE 45: *A shallow cave in McDowell Butte from which rock flowed out at a large angle from the vertical.*

phenomenon for here the material flows out of the cave as before, but the direction of flow is tilted about 45 degrees from the vertical! One can see the effect best by rotating the page appropriately. Plate 46 shows this feature from a different angle. Portions of the caves seen in Plate 44 can also be distinguished just behind it and Barnes Butte is obvious in the background. Notice that the direction of horizontal as defined by the flow in the foreground agrees closely with the level suggested by the bedding features in Barnes Butte, and McDowell Butte slants in this same direction as well. Presumably this implies that the direction of gravity was temporarily

A Fateful Rendezvous

PLATE 46: *The same feature as shown in Plate 45 viewed from another quarter. Barnes Butte is visible in the background.*

and locally altered during the event.

One might notice the well defined "seam" where the flow pattern in the foreground meets the rest of the mass on the right. The junction is smooth and orderly, but it is not perfect. This is one of several explicit junctions that are evident on McDowell Butte, but the topographies on the two sides are not generally so obviously different.

Plate 47 is a view looking east at McDowell Butte which illustrates another example of this apparent twisting of the direction of gravity. Judging by the barely discernible ripple-like patterns on the surfaces, these two prominent rock masses

PLATE 47: *McDowell Butte looking east showing large features with a superficial ripple pattern progressing across the incline.*

also flowed in an unexpected manner, for the motion was far from downhill with respect to our present reference. In fact, the fluid moved directly across the face of a fairly steep incline, although presumably it moved in accord with the direction of gravity which prevailed locally at the time.

Plate 48 is a view looking south at these masses. Notice that the lower flow is nearly flat, as one would expect of a liquid surface, and this level surface defines a horizontal which is consistent with that deduced from the forms in Plate 47. With respect to that level, at least, the two masses flowed directly downhill.

A Fateful Rendezvous

PLATE 48: *An end view of the flowing rock masses seen in Plate 47, looking south. McDowell Road is in the foreground.*

In this grim context we might recall Emil Haury's plaintive comments about his radiocarbon results from the nearby ruins in Snaketown. He found a wholly inexplicable scatter in the results that caused him to question the validity of the method itself. Perhaps we no longer need be surprised. One might imagine that even radiative decay rates could be modified during a paroxism such a this—one where worlds verily fold upon worlds in this bizarre manner. The normal rules of physics simply do not apply in such circumstances.

Would that the Spaniards had sought the New World with more civilized motives in mind. Then the Cities of Cibola would have found their proper place in history, and then also their sudden death would have been a secure fact of record and a challenge to the world of learning for ever after. How much different would have been the development of science and philosophy if that challenge could have been answered in a timely fashion!

One need hardly labor long to decide on a probable cause for that havoc; the impact of a comet is the only possibility that comes to mind*. Although these residues are superficially different from those seen before, the fall of alien dirt is dra-

---

* *One can hardly fail to recognize the strong similarity between this catastrophe and that which destroyed the cities of Sodom and Gomorrah. In Chapter XIX of the Book of Genesis we read in verse 24:*

> " *Then the Lord rained upon Sodom and upon Gomorrah brimstone and fire from the Lord out of heaven.*"

And then the next morning Abraham looked over the scene and saw, according to verse 28:

> " *And he looked toward Sodom and Gomorrah, and toward all the land of the plain, and beheld, and lo, the smoke of the country went up as the smoke of a furnace.*"

If this firey destruction did indeed result from the impact of a comet then an added insight to its consequences may be gleaned from a statement given earlier in Chapter XIII. Describing the time when Abraham and Lot separated to allow greater range for their herds verse 10 reads:

> " *And Lot lifted up his eyes and beheld all the plain of Jordan, that it was well watered everywhere, before the Lord destroyed Sodom and Gomorrah, even as the garden of the Lord, like the land of Egypt, as thou comest into Zoar.*"

Some translators render this verse slightly differently, but the meaning is always the same: The valley of the Jordan was well-watered, even as the valley of the Nile, before the presumed comet struck—which leads one to anticipate that lingering cometary residues may influence the climate and the weather in some inexplicable way.

A Fateful Rendezvous

matically the same. Covering those few square miles with new dirt and rock was but a replay in miniature of what has already been seen. But is there any hope of identifying the specific comet that caused the destruction?

Not with any degree of certainty to be sure. In those days, even more so than today, many comets escaped detection altogether; only the brightest of them were ever seen. But during the interval in question, between 1539 and 1694, when the cities must have been destroyed, records [56] show that 26 comets were observed well enough that their orbits could be subsequently calculated. An examination of those orbits shows that none of them could possibly have struck the earth. However, let us recall that comets sometimes divide into several parts so that on any given orbit several distinct daughters of an original parent may also be present. It is therefore of interest to inquire whether the earth passed close enough to any of those 26 orbits to have collided with a daughter comet that might have been leading it or following along behind.

It happens that only two of those orbits passed within 1,000,000 miles of that of the earth. The comet of 1684 missed by about 790,00 miles while the great comet of 1680 passed less than 250,000 miles from the earth's orbit. When one reflects that any such daughter comets were likely formed during a previous passage around the sun hundreds of years previously, this small discrepancy is no more than trifling. Even supposing that the observations of the great comet itself were precise, a daughter might well have wandered that far astray during all those intervening years.

So the comet of 1680 is easily the most likely candidate to have been the "parent" of the supposed small fragment that destroyed Cibola. Not only did its orbit carry it closest to the path of the earth, but this one was also the largest and therefore perhaps the one most likely to have had fragments associated with it. In fact, by all estimates [32;p.332] that great comet was the most magnificent that the world has ever seen. It passed closest to the earth's orbit on the 21st of November while it was approaching the sun. However, the earth did not arrive at that same point until the 22nd of December*. The fragment that may have struck the earth, then, would have been following nearly the same orbit, but trailing about 31 days behind. As the earth approached that point the parent comet continued on its way and soon became invisible in the glare of the sun. It passed behind the sun, emerged from the other side, and became visible again in the evening about the same time that the earth arrived at its fateful rendezvous.

Now at last we can deduce the probable cause for those 18 feet missing from the bed of the Salt River. To this end one must understand that both the Salt and Verde Rivers drain high mountainous territory to the north and east that accumulates substantial deposits of snow during the winter. By the 22nd of December an appreciable amount of snow could easily have collected, so if a firestorm similar to that which rained fire down upon Wisconsin, Michigan and Illinois had attacked this region then great quantities of snow would have melted immediately. But mid-winter is already a rainy season in that area so the runoff from that suddenly melted snow would have added to the normal flow stemming from the lower elevations, thereby giving rise to extraordinarily severe flooding. Presumably, then, this flood of the millenium can be blamed for the badly eroded condition of the river bed.

It is an interesting sidelight that while all this was going on Father Eusebio Kino was awaiting his assignment overseas at the Jesuit College at Cadiz. Six days after that tragic event, on the 28th of December (1680), he wrote to the Duchess of Aveiro to give his impressions of this great comet that had then become brightly visible again in the evening. Little did he suspect the part that he would play in the drama 14 years

---

* *The details of the calculation leading to this result are deemed to be beyond the scope of this study.*

A Fateful Rendezvous

later—or that some three centuries afterwards we would be reading his mail. But let us now read a portion of the letter he wrote on that occasion and get his firsthand account of that magnificent object [19;p.95]:

"... Already [for the 5th day] at six, seven and eight o'clock, we beheld here a huge comet, which I do not doubt was clearly visible in Madrid, but probably disappeared there beyond the horizon an hour earlier than here. On the 23rd of this month, it was first clearly visible to us who are staying in this college; although some had already detected it three or four days earlier.

" I have no doubt that this is the same comet which many say they saw before sunrise (between four and five A.M.) some four or five weeks ago. They beheld it in the east with its nebulous train pointing westward. ...

" That the comet's own motion or lag displacing it from west to east and at the same time diagonally northward was at the rate of almost four degrees daily, I could observe here on the preceding five days, namely on the 23rd, 24th, 25th, 26th and 27th of this month. Consequently, whereas I calculated that the comet's head on the 24th of this month appeared to us in Cadiz from the occiput of Sagittarius, on the 27th I ascertained that the comet's head had reached the foot of Antinous, so that it seems most likely that in five or six more days it will have ascended to the Dolphin and Aequiculus; and, thus, for several weeks yet, it will enter on a much higher course. We have established that the train of the comet covered some fifty or more degrees, and hence was one of the largest ever seen, extending as it did from the head of Sagittarius to the wing of Cygnus, and hence from the tropic of Capricorn to the tropic Cancer and beyond. ...

" Concerning the distance of this comet from the earth, its size, its exact position and for what regions (especially European) it seems to presage and threaten disaster, I shall strive to make clear on proximate and better occasions. ..."

Both because of its great size, then, and because the earth

passed closest to its orbit, this comet stands alone as the most likely parent of the object that destroyed Cibola and the other communities. In that case the Cities of Cibola, Totonteac, Marata, Acus and the others met their end in the mid-afternoon of December 22 in the year 1680. On that fateful day they passed into oblivion, their cries of anguish lost in the wilderness. There was none to tell the story because no survivors ever reached the Spanish domain where an account could have been preserved. Some might have made their way to New Mexico to be sure, but recall that only four short months previously every white man in that vast region had been either killed or driven out. Not one penman remained to make a permanant record.

Other great civilizations of the past left monuments behind so succeeding generations could marvel at their grandeur, but the residues of Cibola are humble indeed; witness the shambles in Plate 3. No marble columns are to be seen here, but although those cities never acquired the technological trappings of the European nations they surpassed the latter in other qualities of a civilized community. They had the wisdom and the humanity to live together in peace, during fat times and lean, for multiplied centuries. And according to the description left by Fray Marcos they were an industrious, refined and intelligent people who were admired and respected by their neighbors. One feels the tragedy all the more keenly that such an exemplary nation should have passed unmourned and unsung for, truly, to die unmourned is to die indeed.

Chapter 10:

EASTER ISLAND

ONE MIGHT WONDER whether other damage resulted from that cometary encounter of 1680. Certainly history records nothing obvious, and evidently not even rumors of the unusual were circulating afterwards in Mexico. That is, we recall from the final paragraph of Father Kino's letter to the Duchess that he believed comets to be harbingers of disaster (a common belief in those days and justifiably so as it happens). And we recall further that his first act upon arriving in New Spain was to see to the printing of a monograph [53] detailing his observations of the recent comet. Therefore, if suggestive rumors had been circulating at the time he would certainly have taken notice and made mention of them in his book. But he was silent on the topic.

As we now understand, relatively tranquil peripheral phenomena can also result from cometary encounter—Fortean-like effects that should be noteworthy even if not overly severe. A search of contempory Mexican records with this thought in mind might well bear interesting fruits, but we had best leave this task to those who are properly equipped for it and direct our attentions elsewhere. In fact, suggestive clues are already widely known in another context so let us take a few moments to consider them.

Forty-one years after that fateful event, on Easter Sunday of 1722, a Dutch sea Captain by the name of Jacob Roggeveen discovered an island in the south Pacific which harbored a multi-faceted mystery—one that has held the world in thrall to this day. Having dubbed his find, "Easter Island", Roggeveen landed a scouting party to investigate, and he noted many huge statues standing around and about. At first he marveled that those primitive people could have erected such great monuments with no heavy timbers or cordage at hand, but upon superficial examination he concluded that the statues were modeled in clay. In point of fact they were carved out of volcanic tuff at a remote quarry and were somehow moved over rudimentary paths to stations all over the island.

As it happens factual information from those ancient times is scarce because more than a century passed after the time of Roggeveen before serious effort was made to reconstruct the history of the island and discover the meaning of the monuments. By then it was too late to obtain definite information because several tragic circumstances combined to confuse the people's links with their past. On the one hand Peruvian slavers raided the island around 1860 and took many away to work guano deposits in the Chincha Islands. And then on the other hand, the natives were then constantly engaged in vicious inter-clan wars that further decreased the population and degraded their ancient ties. Negotiations with Peru eventually caused a few of those slaves to be returned, but the latter brought deadly smallpox back to the island which reduced the population even more. Thus did the past become ever more cloudy so that only a few vague shadows remain today. Nevertheless, let us review these shadows as best we can.

We might begin with a few words about the island itself. It was formed from the outpourings of three main volcanos, one of which provided the material for those great statues. This one is called Rano Raraku, and its walls are of a fine-grained tuff very suitable for carving. Plate 49 shows a few of the

Easter Island

PLATE 49: *A few of Easter Island's giant statues standing on the slope of the quarry volcano Rano Raraku.*

PLATE 50: *A giant statue with top-knot restored to place in modern times.* PARIS MATCH *photo/Saulnier.*

Easter Island

statues—called *moai* by the natives—standing on the slope of Rano Raraku. These were still in the finishing stages and had not yet been taken to other sites for permanent mounting. Those that had been established at permanent sites were, without exception, deliberately toppled over during the inter-clan strife. Usually the tipping was carefully arranged so the heads would break off when they fell.

Plate 50 is a photograph of one of these statues celebrating the occasion of being restored to its original position. Its majestic size is made all the more evident by the crowd of people standing before it. Note that it wears a hat, as many of them did. These top-knots, as they are called, were carved from a deep-red tuff quarried from a small crater about four miles removed from Rano Raraku. They were balanced on the heads of statues after the latter were finally in place upon their foundations. Of course the means by which these stones were manipulated so adroitly has remained a mystery to this day. About six hundred of the statues were carved in all. They differed in size and in minor detail, but with only one exception they all seem to have been fashioned after the same model.

Modern attempts to explain the mode of transport of these massive statues have all taken for granted that the large timbers which would have been needed for levers, rollers and derricks* were readily at hand, but all signs indicate that the island was devoid of large plants in those days. There were certainly none when Roggeveen visited, and indeed the very life style of the ancient people testifies to the absence of load-

---

* *Some writers have supposed that these simple tools alone would have been sufficient for the task, but they overlook the fact that this tuff has very little strength in shear—as is obvious from the fact that vandals in former times were able to break off the heads of statues so easily. In fact to move those huge loads safely by customary means the larger statues, at least, would need to be housed in an elaborate cradle, so designed as to support the load uniformly.*

bearing wood even from the earliest times. Plate 51 shows one common form of dwelling from that period-called a "boathouse" today because of its characteristic shape. To some extent this is a conceptual model because nothing remains of the original structures, although some foundations survive so there can be no doubt about the general form; one of these is seen in Plate 52. It consises of a number of easily manageable stones set in the ground and fitted end to end. Then socket holes were drilled at intervals to receive the rib-like elements which were a tall grass similar to bamboo. Natural caves in the volcanic rocks were also utilized for shelter. One must wonder

PLATE 51: *Model of an ancient boat house photographed in the Father Sebastian Englert Museum on Easter Island.*

Easter Island

why a people with the vision to conceive and execute those grand monuments would have lived in such cramped and primitive quarters unless compelled to do so—King and drudge alike—by ultimate necessity. In fact, nowhere on the island is there any direct evidence for the ancient use of structural wood or strong cordage.

But the statues are only part of the riddle; the other part concerns the people who produced them. For it appears that some sudden, long-forgotten tragedy effectively destroyed the statue-building culture sometime during the latter half of the 17th century. One cannot fail to notice that this includes the

Plate 52: *Original foundation of an ancient boat house. It lies very near to the restored statue in Plate 50.*

time of the comet so the nature of that tragedy is a matter of special interest. As already noted, the past is very dim here, but careful studies of the ancient traditions have recovered the following basic information:

Easter Island was formerly populated by two races of people who came to be called the "Slender People" (Hanau Momoko) and the "Stout People" (Hanau Eepe). The Slender People arrived first, coming by canoe from a land called "Hiva", and their leader became the first king in the royal line. The date of that landing is not known, but it seems to have taken place at least several centuries before the time of the comet. The Stout People came sometime afterwards; they were men only so they took wives for themselves from the other tribe. These Stout People were a minority, but they appear to have been somewhat more clever than the others because they soon came to be the foremen and effective rulers of the island.

As it happened, the King occupied a most unusual position in the community; in fact he hardly ruled at all in the normal sense. Custom has it that members of the direct royal line possessed a certain power called "mana" that worked for the general good of the people. Whatever may have been the basis for this idea, it was the foundation for the whole culture. According to the natives it was an entirely impersonal power that had its seat within the King's head. As a consequence of this belief, skulls of past kings were sometimes stolen from their graves in hope that whatever mana remained might work locally to the advantage of the thief or his clan. The King's only duties, then, were to preside over certain ceremonies and to propagate the kingly line so this mana could be passed on to his first-born son.

Years ago, when the natives were first asked how the statues had been moved from the quarries to other parts of the island, their invariable reply was that they moved, in effect, all by themselves by means of the mana. They explained that mana no longer existed on the island—that it died with the last

Easter Island

king of the line in about 1870. But the statues had stopped moving long before. In fact, most had already been tipped over by that time, and they could not be righted again.

The quarry mountain, Rano Raraku, constitutes a kind of "window" looking out upon those former times. Through it one can easily see that a great many workmen took part in carving the monuments, and it is equally clear that work stopped suddenly, as if in a single day. In fact more than two hundred of the partially made statues are still to be seen there, some very nearly finished while others were only just begun. Even those that were in transport along the way stood fast and moved no further. If work had trailed off deliberately then certainly the inventory of statues in progress would have been smaller or even altogether nil.

It is worth noting here that despite this sudden and unexpected work stoppage, with many statues stopped in transport, no trace of any transporting apparatus survives—either along the wayside or upon the quarry mountain. Furthermore, statues had been carved—and were being carved both inside and outside of the volcanic cone at sites accessible only with a sure foot, and even at those places no ramps or engineering emplacements are to be found, nor any sign that such ever existed.

Tradition accounts for the work stoppage in this way [31;p.130]: Upon an occasion, the Stout People ordered the Slender People to pick up all the stones on the island and throw them into the sea. The Slender People, not wishing to be ordered about any longer, rebelled, and the decisive battle took place at a kind of trench, called the Poike Ditch, which separated their respective domains. The Stout People, having got wind of the rebellion retreated behind this trench and threw all kinds of burnable debris into it; their plan was to ignite it while the Slender People were crossing the ditch to attack them. However, by a ruse the Slender People turned the tables and threw the Stout People into the burning ditch instead. Only one lone man from among the Hanau Eepe was permitted to survive—that he might have descendents.

Father Sebastian Englert, for a long time a priest on the island and a student of the native traditions, was able to reconstruct the approximate date of the rebellion by a careful study of that man's progeny. More exactly, he placed the birth of that lone survivor somewhere in the latter half of the 17th century [31;p.134]. This is only approximate, of course, but it is significant because other evidence points to that same time period. In particular, excavations in the Poike Ditch were later carried out by a Norwegian expedition under Thor Heyerdahl, and a charcoal sample was taken from what appeared to be the remains of a great fire. Carlyle Smith interpreted the subsequent radiocarbon assay in these words [76;p.391]:

" This leads to a slight reinterpretation of the legend. The Hanau Eepe may have retreated behind an existing fortification which had been constructed many generations earlier to meet just such an emergency. Our investigations support the legend to the extent that the ditch and mound are undoubtedly man-made, and that a great fire burned in the ditch. No traces of burned human bones were found, but when the length of the ditch is considered the chances of finding such remains were remote.

" The date of *ca.* 1680 for the great fire, and the absence of the *mataa*, or obsidian spear head used in warfare, suggest that the war between the Hanau Eepe and the Hanau Momoko marked the end of the Middle Period, when the *mataa* seems to have been unknown or rare, and the beginning the Late Period when it was common."

Now such radiocarbon dates are only deemed accurate to within a hundred years so this exact agreement with expectation is fortuitous, but it is undeniably reassuring. Evidently a profound change in the social structure took place at about the time of the comet, and several strong clues suggest a direct connection between the two events.

Easter Island

The first is a provocative legend from ancient times recorded by Maziere [57;p.57] that goes like this:

"In the days of Rokoroko He Tau the sky fell.
Fell from above on to the earth.
The people cried out, 'the sky has fallen
  in the days of King Rokoroko He Tau.'
He took hold: he waited a given time. The sky
  returned; it went away and it stayed up there. ..."

This certainly must be considered a remarkable statement. Otherwise wholly devoid of meaning, it is most apt indeed in light of what we now know about matter falling from the sky, stimulated by the close passage of a comet. But there are minor complications.

It happens that there were two of the royal family who were known to bear that name, but only one was actually a king. This one was next to last in the line of succession, and he died as a boy of tuberculosis in the year 1867. The earlier one was a son of a previous king, though he was not heir to the throne; consequently one must interpret the legend slightly. To this end let us hear Métraux tell about that former Rokoroko He Tau; this noted French authority is here discussing the power of mana [61;p.90]:

" This power over nature was concentrated in the eldest son; but sometimes it developed such intensity that it risked becoming the source of numberless evils. The legend of the little prince Rokorokohetau, son of the third wife of King Nga'ara, affords a famous example of this. The case is particularly curious because this king's son had, by birth, no right to royal dignity. His entry into the world was accompanied by wonders such as generally announced the birth of a great chief. Many people were devoured by sharks, and sea beasts appeared on the shore and attacked those who ventured there. Finally, white fowl—hitherto unknown—began to multiply. These miraculous events were manifestations of Rokorokohetau's mana. In the hope of averting these disasters and saving his people, the reigning king had his son taken away and shut up in a cave on Mount Rano-aroi. In vain—because his subjects, convinced of the sanctity of the little chief 'with the diadem of white feathers', refused to carry before the legitimate heir the standards symbolic of royalty. In the end Nga'ara had his son, whose mystic power had such baleful effects, put to death".

These remarks deserve careful attention, but for the present let us merely observe that although this prince was not in line for the throne, nevertheless, owing to the highly unusual circumstances, many of the people acknowledged him as King. Therefore the legend could very well have reference to this one, but the list of kingly successions as reconstructed in modern times poses a more serious problem. For King Nga'ara, the father of that controversial prince, is given as only the third before the legitimate King Rokoroko He Tau. If this were actually correct then to associate the comet with that reference to the falling sky would be stretching a point; the three reigns would have had to span nearly two hundred years—not impossible, but not very likely either. However, let us take those dark days in the last century into account and acknowledge that there may be gaps in the names that survived.

In that case the legend of the falling sky deserves further attention, so with it in mind let us reconsider that strange order of the Stout People—to pick up all the stones on the island and throw them into the sea. It is scarcely credible as it stands because Easter Island is exceedingly rocky. Indeed it is not much of an exaggeration to say that there would be little left of the island if its stones were all thrown into the sea. But a few specific stones are noteworthy because they are decidedly alien; they are not of volcanic origin at all. These are small, smoothly rounded pebbles, somewhat flattened, typically about two inches across, and they are all to be found on the "front porch" of the Ahu Akivi. Plate 53 is an attempt to display these

Easter Island

PLATE 53: *The Ahu Akivi with its seven restored moai and a multitude of alien pebbles barely visible on the front porch.*

alien stones in place. The large rounded rocks are of volcanic origin and are typical of all the Ahu, but the small stones between the latter, only barely distinguishable in the photograph, are unique to this one.

This multitude of foreign stones, all collected at this one imposing spot, suggests that the legend became somewhat distorted during those bad years. Possibly the original phrasing went something like this: Upon an occasion, the Stout People ordered the Slender People to pick up all the *newly fallen* stones on the island and throw them into the sea, but the Slender People *insisted upon offering them to the gods instead.* In that case the conflict that followed was a religious war, often the most brutal kind of all.

We have seen the next clue already in Plate 51; the giant statues standing on the slope of the quarry mountain Rano Raraku have been buried under upwards of ten feet of earth and debris! Only the heads of those huge statues are to be seen protruding through the overlying mantle. Commentators have come to refer to this material as "rubble" because they suppose it to be ejecta from the quarries, but its volume is unquestionably many times greater than that of all the quarry holes combined. Plate 54 is a view of Rano Raraku showing the vast

PLATE 54: *Showing the extensive piles of earth and its buried statues at Rano Raraku.*

Easter Island

extent of this dirt deposit. In order to gain a finer feeling for its appearance let us attend to Arne Skjölsvold, a member of the Heyerdahl expedition, as he describes it in his own words [75;p.342]:

" The most convincing evidence of the extent of the industry in Rano Raraku in ancient times is found on the outer edge of the southern part of the mountain. Here are long and impressive piles of grass-covered quarry rubble along the side and foot of the mountain. It was immediately apparent that these characteristic features in the local terrain were not natural formations, but the work of men. Scattered upon and between these earthworks are about fifty gigantic stone statues, most of them buried up to the neck in earth and rubble. ...

" There are good reasons for believing that quite a large number of statues are hidden below the surface. This applies particularly to the area at the foot of Rano Raraku, where enormous quantities of rubble from the quarries probably cover a large number. Both our own and Routledge's excavations have shown that there are good reasons for such an assumption."

He goes on to give a few measurements of those mounds and then he concludes as follows:

" It is impossible to visualize how the volcano looked originally. The quarrying activity and the enormous quantity of rubble have completely altered the local topography."

Now is there any conceivable reason why those workmen should have gone to the trouble of quarrying and carving those huge statues and then proceded to bury them again? To ask the question is to answer it. Notice that the writer describes this material as "earth and rubble"; later on he gave more details. The members of the expedition excavated a trench, and then they indicated on a diagram the various textures that were visible in its wall. Distributed irregularly (not in order from top to bottom) the textures were given as:

Black soil, stone picks, much stone debris
Light sandy soil, stone picks, much stone debris
Sandy soil, clay, stone debris, stone picks
Black soil, clay, stone debris, stone picks
Coarse stone debris, Etc.

Without a doubt there must have been much debris left over from the carving operations, which must have been disposed of somewhere in the vicinity, but the black soil, the light sandy soil and the clay found here are decidedly alien, and they must account for most of the enormous volume of this deposit.

PLATE 55: *Giant statues buried in the rubble at Rano Raraku showing pronounced tilting.*

Easter Island

It is interesting to note in Plate 54 that traces of the material are to be seen even at the very top of the mountain.

As an added note, upon careful examination of the base of the hill, just to the right of center in the picture, one can see what would have been the largest of all of the statues; although not quite finished it would have stood more than 65 feet tall. The task of moving this huge load safely would be prodigious even today with the best of modern equipment because of the weakness of the tuff in shear, but apparently the ancient workmen foresaw no problem with it because they had already finished the topknot that was to stand upon its head. That huge red cylinder still stands near the rim of the volcano Puna Pau where it was carved.

Apparently this was a scene of tornado-like fury, even as was reported during the holocaust in Chicago, for note in Plate 49, and again in Plate 55, that during the process of being buried many of the statues were also tilted. In fact, without the accumulating debris for support they would have fallen over—as many of them did according to the report of Arne Sjkölsvold, cited above. Presumably one would not think to trace this tilting to the vandalism of later years because if great force had been applied to the statues after they were already buried then the heads would have broken off. Of course all of these evidences, individually and in combination, are readily consistent with our conclusion that the Island was struck by a small comet during the latter half of the seventeenth century, and weighing probabilities it seems most reasonable that it was indeed a piece of the same comet as ravaged Cibola. Looking back on the horrors of that fateful day one can but feel relieved that such catastrophic encounters take place only rarely.

Recall that we were able to formulate our new picture of comets by granting a measure of truth to various seemingly fanciful legends from antiquity. We accepted the thesis that the stories would not have survived the first telling if people of the time had not understood their factual basis. Here on Easter Island we found that same principle holding true. Though distorted and incomplete the legends told the simple truth about events so unfamiliar as to defy description in prosaic terms. In the pages to follow we shall examine still other unlikely legends from old and likewise hope to recover long-lost grains of truth hidden within them.

Chapter 11:

ATLANTIS, ET CETERA

IN THE LIGHT of what we have discovered about the way comets interact with planets it is surely only a small step to identify the ring system of Saturn as the residue of just such an encounter. Presumably the eye of the captured comet, at rest with respect to the rotating planet, was situated so far above the surface that the emerging materials condensed into near synchronous orbits where they remained aloft instead of falling into the atmosphere below. If we denote by r_e the equatorial radius of Saturn then one can show that the synchronous orbit would lie approximately $0.81 r_e$ above the visible surface of the planet, whereas the ring system extends between $0.17 r_e$ and $1.27 r_e$ above the surface; it therefore includes the synchronous point. The relatively great width of the ring system suggests that the materials were ejected out from the eye in all directions with a substantial range in velocities. Perhaps this came about because the comet decayed rapidly, the solids being propelled away by gases that were emitted at the same time. In that case one might expect that the decaying comet would have been visible from the earth.

Confirmation is clear and immediate since we have the word of Diodorus Siculus that in his day (during the first century B.C.) Saturn was the most prominent of the planets.

Cardona [22] quotes him as follows:

> " ... But above all in importance, they [the Chaldeans] say, is the study of the influence of the five stars known as planets. The one named Cronus [i.e. Saturn] by the Greeks is the most conspicuous. ... They [the Chaldeans] call it the star of Helios."

Helios is, of course, the Greek word for the sun. Cardona, who has made a special study of legends relating to this planet, reproduces many instances from antiquity where Saturn was celebrated as by far the brightest of the planets, whereas today, viewed from earth, it is the least bright of the five visible to the unaided eye. Thus, Saturn gave every outward appearance, as we would now understand it, of having captured a large comet during historical times*. Of necessity appearing as only a small point of light it was nevertheless extremely bright and acquired the epithet 'Son of the Sun'—suggesting that the planet was easily visible in the daytime. In fact, according to Cardona, it and the sun were often called by the same name, and not in Greece only but in Mesopotamia and India as well.

He then goes on to cite many instances from antiquity, almost to the point of monotony, which *identify Saturn with time*. In the Greek language this identity

$$\text{Kronos (Saturn)} = \text{Chronos (time)}$$

is usually put down as philological naivete, but the equality is often spelled out explicity—and not merely in the Greek lan-

---

* *According to our findings in Chapter 8, at that very same time Venus also harbored a resident comet and yet, judging by the reports, the most noteworthy feature of that comet was not its brightness but rather its cometary appearance, beard and all. One might suppose that the brilliant eye of the comet on Venus, lying closer to the surface of the planet, became obscured by the dust-laden atmosphere above, whereas the eye of the comet on Saturn was situated well above the atmosphere and was therefore visible from afar.*

Atlantis, et Cetera

guage, according to Cardona, but in the Sanskrit, Indian, Egyptian and Persian also. Our view, tracing Saturn's former brilliance to the eye of a comet situated near its equatorial plane, gives an easy explanation for this seemingly absurd equation. Namely, seen from the earth that brilliance would blink on and off periodically as the planet would occult the rotating incandescent eye. Out of a period of ten hours and fourteen minutes (the rotational period of the planet) the eye might be invisible for as long as an hour and fifty-four minutes, depending on the position of the earth with respect to Saturn's equatorial plane. This regular blinking would measure out time, even as does the beating of a pendulum.

We might take a moment to note in passing that this blinking of Saturn easily resolves still another remarkable mystery from olden times, only recently brought to light. This puzzle involves a map produced by one Piri Re'is, an admiral in the Turkish navy, in the year 1513. According to its title the map originally charted all the known world, but in the intervening years the brittle parchment cracked and much of it was lost. What remains includes the Iberian Peninsula, the western hub of Africa, part of Antarctica, and the eastern coasts of the Americas as far north as Cuba. Writing on the map itself Piri Re'is expained that he had compiled the data from ancient charts, some dating back to the time of Alexander the Great.

The map bears little resemblance to anything produced in recent times; most notably lacking is a grid defining latitudes and longitudes. In their place five wind roses are laid out at nominally equal intervals along an arc in the Atlantic ocean, from each of which up to 32 lines radiate outward. The center of that arc lies off the page in a portion of the map now lost and no numbers are anywhere given. Overall it appears to be an impressionistic hodge-podge drawn by some early artist as a mere ornament.

But Charles Hapgood and his students and Keene State College undertook to deduce the system that had been used to draw it. Their efforts read like a detective story [42]. Much painstaking labor was required, but eventually they deduced the scale of measurement, the meaning of the wind roses, the center of that arc (which defined the prime meridian), and the trigonometric projection that was used to plot the spherical earth onto a plane. With these points all understood the coordinates of identifiable features on the map could be read with surprising accuracy; in his Table 1 Hapgood listed 65 identifiable sites in the western hemisphere, Antarctica* and the south Atlantic and compared their positions as plotted with their true positions as known today. Cuba alone is considerably out of place with its 12 identifiable points being located on average 5.2 degrees too far west. But the average error in longitude of the remaining 53 sites is only 1.4 degrees!

Now one should understand that establishing the relative longitude of points beyond the range of direct communication reduces to *measuring the difference between their local times*. Before radio communication this involved measuring the hour angle of the sun at the given site and then subtracting the hour angle prevailing simultaneously at the prime meridian—this latter angle being calculated from the reading of a well-regulated clock that had been set to read local time at that reference meridian. Before development of the sea-going chronometer the longitude of points in the western hemisphere with respect to Greenwich could therefore only be guessed. That the Piri Re'is map should have plotted western sites with such precision in the year 1513 therefore constitutes a fundamental mystery that prompted Hapgood to remark [42;p. 49]

" ... By default of any alternative, we seem forced to ascribe the origin of this part of the map to a pre-Hellenic people—not to

---

* *It is interesting to note that the contour of Antarctica as shown on this map closely agrees with the contour of the actual land mass—a contour which can today be determined only by seismic soundings through the ice cap.*

Atlantis, et Cetera

Renaissance or Midieval cartographers, and not to Arabs, who were just as badly off as everybody else with respect to longitude, and not to the Greeks either. The trigonometry of the projection (or rather its information on the size of the earth) suggests the work of Alexandrian geographers, but the evident knowledge of longitude implies a people unknown to us, a nation of seafarers, with instruments for finding longitude undreamed of by the Greeks, and, so far as we know, not possessed by the Phoenicians either."

Of course, a plausible solution to the mystery is now easily given. That instrument was merely the planet Saturn blinking out time independently of the rotating earth*.

Further corroboration for this picture can be gleaned from data returned from the Voyager 2 space probe as it passed through the ring system of Saturn. At this unique time some of the instruments on board operated erratically—or failed to operate altogether, but after the craft had safely passed through that critical region they recovered their normal function and

---

* Note that this "clock" would gain or lose time in response to the changing separation between the two planets. That is, a change in separation by 186,000 miles would give rise to an offset of one second. This offset could accumulate a substantial error in response to the changing angle between Saturn and the sun on the celestial sphere. Since the planet was probably visible in the daytime the range of variation would have amounted to nearly 16 minutes. An even larger deviation would have resulted from the changing elongation of the earth from the sun, as seen from Saturn—which can be as much as six degrees. The resulting change in phase of the blink would have alternately advanced or retarded the clock by as much as 24 minutes from its "setting" with Saturn in opposition. Moreover, the sun itself fails to keep time uniformly owing to the ellipticity of the earth's orbit. The overall problem would therefore be fairly complex, but except for the group that mapped Cuba the ancient mariners were evidently aware of these deviations and corrected for them successfully, even though they may not have understood their origin clearly.

control was restored. Measurements which had been recorded during that time were then played back and analyzed. David Morrison amplified on the riddle and described the first reception of that data as follows [63]:

> "... Even after the tape recordings had been received, it was still not evident how the failure had occurred. It did not appear to be a discrete event associated with the ring plane, but rather a progressive degradation of the capability of the scan platform to move as directed. ... At the 2 p.m. science meeting, the most spectacular results were presented by Fred Scarf [Principal Investigator for the Voyager Plasma Wave Experiment]. Very close to the time of ring plane crossing, the plasma wave instrument recorded activity a million times the normal energy level. The high frequency of the signal proved that it could not be ordinary plasma waves, but more likely an electrical phenomenon taking place at the spacecraft ... The roaring sound ... on the tape ... sounding almost like a hailstorm striking a tin roof, sent chills down the spines of the seventy-five scientists attending the meeting. ..."

Of course, one cannot pretend to understand this bizarre behavior in detail, but generally speaking he must conclude that a faint residue of that comet still survives, even if it is no longer visible, and that Voyager 2 passed through its active eye on that occasion.

Another remarkable legend from olden times devolves around the island kingdom of Atlantis. Egyptian priests originally told the the story to Solon, and then some two hundred years later Plato relayed the account in his own writings. Let us note what one of those priests had to say about that island nation [29;p.10]:

> "... This power came forth out of the Atlantic Ocean, for in those days the Atlantic was navigable; and there was an island situated in front of the straits which you call the Columns of

Atlantis, et Cetera

Heracles; the island was larger than Libya and Asia put together, and was the way to other islands, and from these islands you might pass through the whole of the opposite continent which surrounded the true ocean; for this sea which is within the Straits of Heracles is only a harbor, having a narrow entrance, but the other is a real sea, and the surrounding land may be most truly called a continent. Now, in the island of Atlantis there was a great and wonderful empire which had rule over the whole island and several others, as well as over parts of the continent; ..."

This is a remarkable statement. Not only does the priest refer to a continent on the other side of the Ocean, more than two thousand years before the time of Columbus, but he also implied that the Atlantic was not navigable in his time. He then went on to describe a war between Atlantis and the ancient Greeks, after which discourse the priest continued as follows:

" ... But afterwards there occurred violent earthquakes and floods, and in a single day and night of rain all your warlike men in a body sunk into the earth, and the island of Atlantis in like manner disappeared, and was sunk beneath the sea. And that is the reason why the sea in those parts is impenetrable and impassable, because there is such a quantity of shallow mud in the way, and this was caused by the subsidence of the island."

This tale has given rise to much speculation about the location of Atlantis and many theories purporting to explain its sinking; in modern times probably those of Donnelly and Velikovsky are the most prominent, but none has been at all satisfactory. Before contemplating the riddle for ourselves let us attend to a few more words from the priest:

" Let me begin by observing, first of all, that nine thousand was the sum of years which had elapsed since the war which was said to have taken place between all those who dwelt outside the Pillars of Heracles and those who dwelt within them; this war I will now describe. Of the combatants on the one side the city of Athens was reported to have been the ruler, and to have directed the contest; the combatants on the other side were led by the kings of the islands of Atlantis, which, as I was saying, once had an extent greater than that of Libya and Asia; and, when afterward sunk by an earthquake, became an impassable barrier of mud to voyagers sailing from hence to the ocean. ..."

Let it be noted that the priest here spoke of the "kings of the *islands* of Atlantis" (in the plural), and furthermore, he mentioned that Atlantis held sway over part of that continent on the other side of the ocean. We find supporting evidence for this statement in an interesting relic found in Ohio early in the last century. Josiah Priest, in his book, *American Antiquities*, recounts the discovery in these words [67;p.125]:

" ... A gentleman who was living near the town of Cincinnati, in 1826, on the upper level, had occasion to sink a well for his accommodation, who persevered in digging to the depth of eighty feet without finding water, but still persisting in the attempt, his workmen found themselves obstructed by a substance which resisted their labor, though evidently not stone. They cleared the surface and sides from the earth bedded around it; when there appeared the stump of a tree, three feet in diameter, and two feet high, which had been cut down with an axe. The blows of the axe were yet visible. It was nearly of the color and apparent character of coal, but had not the friable and fusible quality of that mineral. ..."

Here in this stump we have palpable evidence of that legendary kingdom for assuredly no simple agrarian society would have had any use for timbers three feet in diameter. But a great maritime power would have very specific use for the largest timbers available; namely, they would serve as keels and masts for great ships. Furthermore, the axe-marks on the stump testify that the tree was felled only a relatively short time before being buried, suggesting that Atlantis was de-

Atlantis, et Cetera

stroyed by that same event as gave rise to the drift residues and the loess. In that case the priest's words "... in a single day and night of rain all your warlike men in a body sunk into the earth ... " are pregnant with meaning. They are hardly appropriate to a rain of water, but they could apply precisely to a rain of loessian silt, which is well developed in the Rhine valley especially. That is to say, the armies were buried in it.

However several problems with the priest's account are readily apparent. For one thing, the time of that catastrophe is probably not well given by the passage above—that is, nine thousand years before the time of Solon and the priest. Even Velikovsky argued that the interval was much greater than can be supported by historical evidence. As an alternative he suggested that a zero had been accidently added to the number somewhere along the line, which would make the actual interval closer to 900 years and place the event just shortly before the Exodus. But in that case most probably we should "refine" the number even further and conclude that the upheavals that occurred in Egypt just before the Exodus were direct consequences of the same catastrophe that had just taken place far to the north. Thereby could we date that awesome event somewhere in the vicinity of 1445 B.C.

In the light of our present understanding of the phenomenon, as gleaned from those remarkable flow paterns in the rocks at Papago Park, we can perceive how Atlantis might indeed have sunk beneath the sea while at the same time the Atlantic became inaccessible from the Mediterranian Sea. Namely, we have only to imagine that the earth's gravitational field was temporarily twisted by the event, even as it was at Cibola. Any such warping would have caused local changes in sea level, which could well have left Atlantis submerged while giving rise to a land bridge between Ocean and Sea there at Gibraltar. Of course, any such warping of the field would have been only temporary; the earth's normal figure would return when the cometary effects had abated. That the priest represented the effects as still persisting in his own day, nearly 900 years (as we have reckoned) after the event, gives rise to conceptual problems and requires that we interpret his remarks somewhat. As one alternative, one might suppose that the priest derived his understanding of the world entirely from reading the ancient sources—those written shortly after the event itself, and he was simply not aware that the unnatural conditions of which he spoke no longer prevailed.

In other words, at some time long previously, the cometary residues having faded away, the Ocean retreated from Atlantis and returned to Gibraltar causing the land bridge to disappear. In that case it is tempting to agree with Beaumont [9;*passim*] and identify the British Isles as the site of the former Atlantis. Being followed in turn by Iceland and Greenland on the way to the continent on the other side of the Ocean they answer well to the description given by the priest above.

And yet still more riddles from olden times find ready solutions in this new light. Shifting our focus now to address one of them we note that scholars who are acquainted with the classical writers know that in ancient times the year comprised only 360 days, reflecting a slightly slower rate of rotation of the earth about its axis. During those early days the moon also moved more slowly around the earth so that it required 30 days to complete a circuit instead of the 29.52 which obtains at present. Thereby did 12 lunar periods closely agree with the length of the year. Godfrey Higgins cited many references to this fact in his monumental work, *Anacalypsis,* [46;II,p.316 to 326], and in later years Beaumont [8] and Velikovsky [81] did likewise. Surely we cannot take the time to examine all of these evidences, but we should have room for at least one. In particular Velikovsky cites Thebaut in these words [p.330]:

" ... All the Veda texts speak uniformly and exclusively of a year of 360 days. Passages in which the length of the year is

Atlantis, et Cetera

directly stated are found in all the Brahmanas. ... It is striking that the Vedas nowhere mention an intercalary period, and while repeatedly stating that they year consists of 360 days, nowhere refer to the five or six days that actually are part of the solar year."

Evidently the change to our present year did not happen suddenly. To make the point Velikovksy drew upon certain clay tablets from the royal library at Nineveh that recorded contemporary observations of the change as it was actually taking place [p.348]. They continued over something in excess of a half century, spanning a time from the end of the eighth century, B.C., and the beginning of the seventh. That corresponds to the time of the founding of Rome, the First Olympiad in Greece and the time of Hezakiah in Biblical history, which was just before the second invasion of Sennacherib*. As might be expected modern critics charge simply that the people of ancient times were confused—too ignorant to measure the length of the year accurately. But the supposed confusion was not confined to India only; Higgins and Velikovsky were able to cite memories of a 360-day calendar in essentially every ancient civilization in the world. Almost always the change to a 365-day calendar was effected by adding five or six intercalary days at the end of the year—days that were outside of the year and were considered "unlucky". Apart from the obvious ancillary problem one could therefore already conclude that these extra days resulted from a change in length of the year after a calendar of 360 days had been long established. Again follow-

---

* *In Egypt the first mention of this change in the length of the year is found in records dating from the beginning of the 19th dynasty; none is found from the 18th. This synchronism is, of course, badly out of line with the prevailing popular chronology of Egypt, but it agrees satisfactorily with the modified scheme discussed already in the footnote on Page 193.*

ing earlier writers Higgins urged that the ancients divided the circle into 360 degrees to reflect the average daily advance of the sun throughout the year.

Now one can easily understand that any such change in rotation rate would require an applied torque—a twisting effect applied to the earth by some external agency. This necessary torque is, of course, the very heart of the problem because no plausible mechanism for applying it can be imagined within the normal scope of celestial mechanics. One should also understand that a change in the moon's period does not necessarily require an external force acting on the moon itself. That is, the earth and the moon rotate as a unit around their common center of gravity, so a properly applied external force on the earth could change that period of rotation equally as well. Our agenda, then, will be to examine independent evidence for a change in the earth's angular momentum and then inquire how the required torque might have been applied in the light of our new perception of cometary impact. For the sake of completeness we should first review in advance a few basic concepts with which the reader ought to be familiar.

Firstly, we all understand that the sun, in its annual motion, traces out a great circle on the celestial sphere. The plane of its apparent path, called the *ecliptic*, tilts somewhat with respect to the plane of the equator. The angle of this tilt, called the *obliquity of the ecliptic*, is presently 23° 27′. As a result of this tilt the sun moves alternately north and south of the equator during the course of a year, the points of maximum departure being called the *solstices*. They occur on about the twenty-first of June and the twenty-first of December. Those occasions when the sun crosses the celestial equator are called the *equinoxes*; at such times the sun rises exactly at the eastern point on the horizon.

Now the earth in revolving about the sun constitutes a kind of gyroscope, wherein the axis of revolution passes through the center of mass of the earth-sun pair and is perpen-

Atlantis, et Cetera

dicular to the ecliptic. Likewise the earth itself constitutes a gyroscope rotating about its north-south axis. Of these two the first is far more stable than the second so from practical considerations a change in the number of days in the year probably would imply a change in the length of the day, that is, a change in the earth's rate of rotation about its axis.

Next, one should recognize that an external torque randomly applied would change both the rate of rotation and the tilt of the axis—the obliquity of the ecliptic. So to confirm that the earth has indeed experienced such an external twisting it will suffice to show that it has suffered a substantial change in obliquity. In this context, one should note that the axes of both of the above mentioned gyroscopes change direction slowly in response to torques stemming from the gravitational pull of the moon and other planets on the earth. At present the obliquity is *decreasing* at a rate of about 0.78 minutes of arc per century, but of course these torques affect primarily the obliquity and also, being essentially uniform, do not satisfy our present need for a really substantial torque applied during only a limited period of time.

But a remarkable means for confirming just such a torque is to be found in the ancient ruin known today as Tiahuanaco. High in the Bolivian Andes it has long been of special interest because of its intricately precise megalithic craftsmanship. But the deeper riddle goes beyond the stonework itself, impressive though that may be. Figure 8 is a crude diagram of one of the structures there known as the *Kalasasaya*. The original outer wall enclosed a grand courtyard some 440 feet between the east and west wall and somewhat less between the north and south. It was made largely of huge stones brought from afar and intricately fitted together. Sometime later it was partially filled in with dirt, and another structure was built upon the resulting platform as shown in the figure. This newer structure is, however, made from much smaller stones which do not fit so neatly together.

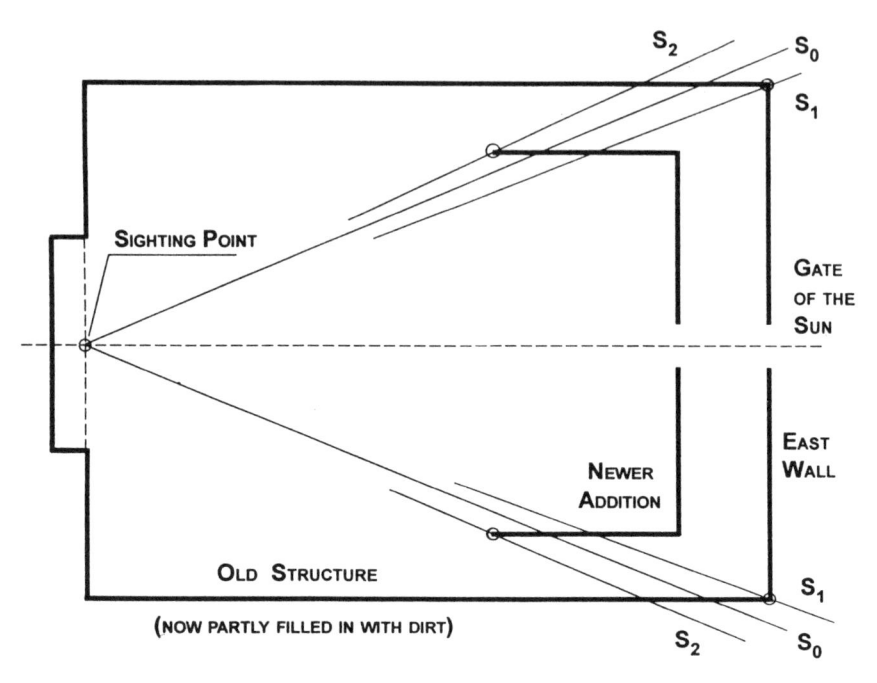

FIGURE 8: *Conceptual plan of the Kalasasaya structure at Tiahuanaco showing the important alignments at the solstices. The four sighting circles on the right represent stones standing up from the wall. Other such stones standing up from the wall may also represent significant sighting points, but they need not concern us here. One should note that the spacing between the sighting directions has been exaggerated for the sake of clarity; thereby are the other proportions distorted as well.*

From the observation point in the niche of the west wall the equinoctial sun may be seen to rise through the "Gate of the Sun" directly to the east, giving a strong hint that a ritual celebration of the rising sun was the central object of this building. Our attention centers on the two lines labeled S_0 which show the position of the rising sun at the solstices. Today they align with nothing in particular, but since the obliquity was greater in the past one can estimate a time when

Atlantis, et Cetera

the solsticial sun rose exactly over the markers at the ends of the new wall—as shown by the lines S_2; presumably that condition prevailed when the structure was built. Let it be said, however, that the date derived from this condition is considerably in doubt because one cannot be certain of the exact criterion the designers used to define the alignment. Nevertheless, according to Steede [77] the new wall was probably erected between 2000 and 3000 years ago; of course, this includes the period during which the change in the year took place.

But now let us consider the rising solsticial sun in relation to the corner stones of the old structure. According to Hancock [41;p.79] it would align to those markers if the obliquity were

$$e = 23° 9'$$

which is 18´ less than prevails today and is not due to be realized until about 4200 A.D., 2200 years into the future! Consequently most academicians regard this near-alignment as a simple coincidence and attribute no significance to it whatever. But Arthur Posnansky, who first studied it in detail more than sixty years ago [65], cited earlier calculations which suggested that the obliquity actually oscillates, during a cycle of 41,000 years, between 22.1° and 24.5°. According to this calculation the value of 23° 9´ would have prevailed somewhere in the vicinity of 15,000 B.C.

This writer is in no position to comment on that calculation, but an age of 17,000 years seems unlikely for any number of reasons. More plausible is the alternative that the original structure was indeed built when the obliquity was 23° 9´, and then some earth-shaking catastrophe caused *an increase* in the tilt by some 38 minutes of arc, along with an increase in the rate of rotation of the earth. Shortly afterwards, presumably, that same race of people—people having the same compelling motive for celebrating the rising sun at the solstices, partly filled in the old courtyard and built the new addition.

In order to account for the supposed change in rotation rate we would need a fractional change in angular momentum of the earth by about 1.5 per cent. Likewise, tilting the axis by those 38 minutes of arc would involve a change of about 1.1 percent, so the two changes are closely comparable; a mechanism that could cause one might well cause the other at the same time.

One can easily test this picture against the simplest possible kind a event, namely the impact of a nominal cometary nucleus with the earth. Let us suppose, for example, that the nucleus were spherical and, say, eight miles in diameter. This would be an even thousandth of the earth's diameter and comparable in size to the nucleus of Halley's comet as observed by the space probe, Giotto. Then it follows from the most fundamental principles of energy conservation that, in falling from the remote reaches of space toward the sun and the earth, this object would have acquired a final velocity of about 27 miles per second just before impact. Moreover, because the earth's own velocity in orbit about the sun amounts to some 18 miles per second we could possibly have a *relative velocity* at impact as high as 45 miles per second. Furthermore, to imagine the best possible case, let us suppose that the impacting mass struck tangentially at the equator travelling directly to the east so as to be maximally effective in speeding up the rotation. Assuming also that its density were equal to that of the earth we find that even in this exaggerated case the resulting change in angular momentum would still be too small by a factor of some 15,000.

Consequently if there had been such an interaction, and if it was indeed caused by the comet, then the mechanism must have been beyond the norms of mechanics, and we have already seen a hint of a plausible explanation in Chapter 9. Namely, we saw that during the destruction of Cibola the *direction of gravity was skewed*, at least long enough for those odd flow patterns to form on the Papago Buttes. Now we find in

Atlantis, et Cetera

Appendix A good reason to believe that approximately half of our observed gravitational field arises from other worlds overlying the earth along that other dimension of space. A local bend in *their* contribution to the prevailing field would therefore appear as a net *transverse external force* acting on one side of the earth. Such a torque would have changed the earth's angular momentum, and an equal and presumably an opposite torque would have been exerted by the earth on those other planets. In reciprocal fashion it would follow that the earth could conceivably suffer spontaneous changes in rotation rate or tilt in reaction to alien catastrophic events that would be otherwise invisible to us here on earth.

Recall that Velikovsky deduced that the disruption continued over a period of some sixty-years, and this too is agreeable to our model for we have seen other cases where phenomena associated with cometary impact persisted for many years. It is interesting to speculate that this may have been the event that melted an existing ice cover of Antarctica so that navigators of a later time were able to map its outline acurately.

Remarkably enough the megalithic monument at Stonehenge adds gratifying corroboration for this tilting even if the fact has not been commonly recognized before. One of its principal features is the so-called heel stone which, viewed from the center of that circle of stones, *almost* marks the position of sunrise at the summer solstice. One lingering question has been the "sight picture" that the builders intended to define the event. Since the top of the heel stone appears flush with the distant skyline it seems most plausible that this should mark the spot where the first flash of the rising sun would appear at the solstice. However, even today the sun rises well north of this spot, and in olden times presumably it rose even further to the north. Of course, we have seen all this before at Tiahuanaco.

Using published figures for the relevant measurements one can calculate the obliquity that would be required for the first flash to appear exactly over the heel stone. Hawkins cites Sir Normal Lockyer's measurement of the azimuth for that alignment as 51.23° east of north. Then he states that the distant skyline lies 0.6° above the true horizon and that the vertical offset of the rising sun due to refraction is 0.47° [44;p170]. And finally we know that Stonehenge lies 51.17° north of the equator. Then one can show by simple trigonometry that the ideal rising would occur if the obliquity were

$$e = 23° 2'$$

However it is interesting to note that Hawkins determined the azimuth of the heel stone for himself by means of cine photographs of the rising sun at the solstice. His finding differed from Lockyer's by 0.15°, but remarkably enough he failed to state the sense of the discrepency. Let us guess that the better value is 0.15° *less* than Lockyer's figure, namely, 51.08°. Then that same calculation gives

$$e = 23° 7'$$

which agrees nicely with the obliquity implied by the Kalasasaya. Taking these figures at face value one would conclude that Stonehenge was built about 250 years after the old structure at Tiahuanaco.

Based largely on radiocarbon assays current wisdom places the beginnings of Stonehenge at about 3000 years B.C. However, if Britain was truly the site of the ancient Atlantis then those assays must be deceptive. After being submerged in the sea some considerable time must have passed before the land could have supported a flourishing population again. Therefore one might guess that the earliest possible date for the construction would be in the vicinity of 1300 B.C. Likewise we know that it was built before the earth-tilting event that started in about 800 B.C. Consequently, one would have to place the beginnings of Stonehenge at nominally 1050 B.C., with an uncertaintly of perhaps 250 years.

Chapter 12:

THE GREAT FLOOD

ONE CAN HARDLY UTTER the word 'catastrophe' without bringing to mind that Great Deluge described in the Book of Genesis. In former times the fact of that flood was regarded as self-evident because residues of its victims could be seen in rocks all over the world—even on the tallest mountains. We recall that the catastrophe is described in The Bible as an outright miracle, wrought by the Hand of God as punishment for the sins of mankind. But as we saw earlier, in keeping with the spirit of Enlightenment during the Eighteenth Century, avant-garde thinkers were determined to account for the face of the earth in prosaic natural terms. Therefore, putting aside any thought of miracles, they imagined that the rocks in question developed gradually as sedimentary deposits within which the various animals came to be entombed during the normal course of events. Then as the ages multiplied these layers eventually turned to stone and silica replaced calcite in the bones. In a nutshell, this is how scientists explain the fossiliferous rocks in keeping with their Principle of Uniformity*.

But in point of simple fact this picture does not conform to that fundamental principle because the transition of silt or

---

* *This principle is explained at length on pages 149 and 150.*

mud to stone is not supported by either physics or chemistry—and neither is the replacement of calcite by silica. And furthermore, we know from simple observation that the bones of dead animals do not form fossils—least of all articulated fossils. Skeletal bones in the wild are typically torn asunder by predators, and after being denuded they weather away in a very few years. Bones in nature as much as, say, a dozen years old must be considered a rarity—and all the more so in an aquatic environment.

In more recent times, the failures of Uniformitarian geology coming into focus, many writers have attempted to explain that great deluge in more-or-less natural terms. One popular view supposes that a canopy of water once enveloped the earth at high altitude. Then, when conditions changed somehow, and the suspending mechanism failed, the water rained down upon the earth causing the flood. Another view traces that great supply of water to the depths of the earth; presumably it burst forth when the internal pressure increased sufficiently.

However, neither of these pictures, nor any conceivable variation of them, can satisfy the need, because the formation of those rocky strata and the fossils within them do not conform to standard physics and chemistry. Therefore, some transcendental mechanism must have been at the root of it all, and fortunately, given our newly found appreciation for the extended scope of nature, we can now contemplate such processes without undue strain.

In this added light it is surely a small step to recognize that the fossil-bearing rocks resulted as Fortean-like side effects of an especially violent cometary collision. These rocks are somewhat different *in form* from the Fortean falls observed previously, but they are not so different *in kind*, and in that case both the rocks and the organisms captured within them could be entirely alien to our world. This would require that the organisms were *chemically modified—fossilized during the transition itself*, and the rocky matrix was evidently in a fluid

The Great Flood

state when it fell. From the footprints of animals sometimes found impressed into such rocks one can conclude that they congealed fairly slowly, preserving the consistancy of a stiff mud for a considerable time. On the other hand, since the prints suffered little damage from subsequent rains, apparently that time must be measured in hours or days at the most. When animals also were involved in such a fall their usual fate was to be caught up in the mass, then to be preserved as fossils, but evidently on some occasions a few fell independently and survived—at least long enough to wander about and leave tracks. In falling should they not have been dashed to the ground and killed outright? One might well so expect, but perhaps he should not be overly dogmatic about processes in this shadowy realm—especially since we have already seen that comets manifest very little mass on impact and survive intact despite their great speed.

As just noted, this picture requires that most victims of such events were fossilized during the fall itself. Such a process is hardly accessible to common reason because it must have involved a transmutation of the elements making up the body. The remarkable specimen shown in Plates 56 and 57 illustrates just such a transmutation, and it also gives a hint of the rate at which the process occurred. Here the jellyfish-like creature was transformed even in the act of swimming, and the water in which it swam was transformed as well! So bizarre is the spectacle that one can face up to it only with difficulty though he hold the tell-tale evidence in his hand. Probably another example of this kind of transmutation is to be seen in the stoney trees of Petrified Forest National Monument in northern Arizona. Renowned throughout the world they are of a flinty hardness and retain the original structures and markings of the living plants intact. According to our new understanding these logs must have fallen exactly as we find them today; no long-term burial stage and subsequent unearthing was ever involved. As one contemplates those logs and the petrified

PLATE 56: *Fossilized organism which may have been similar to a small jellyfish. The animal, and even the water in which it swam, seem to have been transmuted instantly to stone. The pattern has not been artificially enhanced in any way.*

Picked up in northwestern Arizona this specimen probably does not stem from the normal geologic column but may be of a more recent origin, perhaps even a product of the event of 1680. Here viewed from the top it appears to be a rock worn smooth by tumbling, as for example in a river bed, but the bottom displays an entirely different character. The underside is approximately flat, whitish, and it has a molded texture, suggesting that the rock came to rest while in a plastic state. From this aspect it appears to be in its original, pristine condition and shows no evidence of any erosion whatever.

The Great Flood

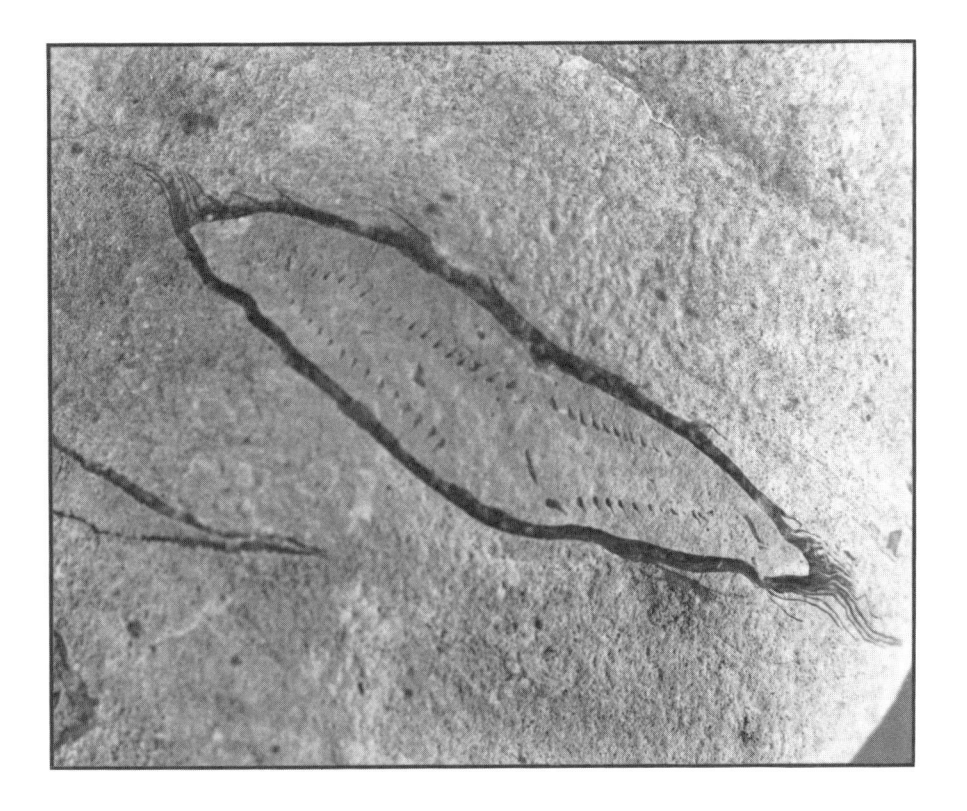

PLATE 57: *A closer view of the specimen in Plate 56. In this section the sample measures about 1.5 inches across, exclusive of cilia.*

jellyfish, with the grotesquely warped rocks in Papago Park also in mind, he must abandon any thought of prescribing limits to the wonders that Nature might work in these extreme circumstances. In this light the folly of the Uniformity Principle becomes all the more apparent.

Now we have already seen that the deposits presently identified with ice ages actually resulted from a cometary collision in the vicinity of 1445 B.C.; these represent most of the record of what is called the Quaternary Period. Then presumably if the Noachian flood actually occurred its residues would be found in the next lower layer of the geological sequence—in rocks of the so-called Tertiary Period. But should they be found in Tertiary rocks only? Even though this layer can be distinguished from earlier deposits by its fossil content, and usually overlies the earlier ones, some instances have been found where the sequence is out of order—the so-called overthrust faults. Billings describes these unexpected features as follows (emphasis added) [12;p. 184]:

" Overthrusts are spectacular geological features along which large masses of rock are displaced great distances. An overthrust may be defined as thrust fault with an initial dip of 10 degrees or less and a *net slip that is measured in miles.*"

Namely, such great masses of rock are thought to have been physically moved over great distances by some unimaginable agency simply because their fossil content departs from the expected sequence. Certainly these anomalies are not faults in the customary sense, and they have nothing to do with thrusting. In the light of our new understanding one would have to conclude that those individual Fortean-like depositions did not come to an abrupt end, but instead they faded away fitfully, with occasional spurts of activity continuing after other depositions had already started. Because this delayed output sometimes fell upon later depositions already in place one would have to conclude that the overall activity was essentially continuous and that most, if not all, of the fossiliferous rocks were deposited within a very short time following that supposed cometary impact.

Because of their differing fossil content one might conclude that the various phases of deposition actually stemmed from different sources. Interestingly enough, we find in Appendix A that other worlds probably overlie the earth along that other dimension so they might seem the most plausible origin of the alien rocks and life forms. In that case each identifiable

The Great Flood

fossil group, as for example the dinosaurs*, would presumably be representative of one of those other worlds. Of course, we would also have to suppose that the sun had an extension along that other dimension as well so that those worlds would have a source of heat and light. We might observe that this interpretation of the fossiliferous layers removes at once whatever justification there may ever have been for the theory of evolution.

Of all those who have attempted to interpret the earth's rocky layers this writer knows of only one who correctly anticipated some of our present findings and who must therefore be acknowledged at this point. Perhaps the reader will find it reassuring to discover that similar conclusions have been derived from entirely independent evidences, so it will be doubly worth our while to review the work of Comyns Beaumont briefly. We saw earlier (p. 124) that he vigorously rejected the ice age theory as an explanation for the drift and associated deposits. Agreeing with Donnelly he traced these materials to the impact of a comet, and he also identified that event as the origin of the Phaëton myth and the destruction of Atlantis. But unfortunately, even as did Donnelly before him, he had a very faulty concept of comets so his reasoning was strained at times and not entirely persuasive.

Nevertheless, his interpretation of the chalk deposits in northern Europe is of interest in our context. First of all, then, Beaumont noted that the chalk is a massive deposit of almost pure calcium carbonate, formed from the crushed shells of countless small molluscs. He then pointed out that there exists

---

Let us note here in passing that this picture also offers an easy resolution to problems posed by the gigantic size of some of those animals. Briefly stated, various of the dinosaurs were so large that their bones lacked the strength to support their bulk reliably, their muscles had scarcely the strength to provide mobility, and with their 50-foot wingspread some of the winged creatures were so huge that they would have been unable to lift themselves into the air. This problem and a possible resolution are considered at length in Appendix E.

a long line of chalk masses, stretching from the north of Ireland to the Crimea, laid down intermittently in thick parcels in a direction from northwest to southeast. Likewise he identified another line of deposits almost at right angles stretching from southern Sweden to Bordeaux, also intermittent. Then he reasoned as follows [8;p.205]:

> " ... If this chalk accumulated in former terrestrial seas, even if its transition from marine shells to chalk were an evolutionary process, we should have to project a period when two enormous but narrow straits existed, cutting through Ireland, Wales, England, Belgium, Germany and Russia on the one hand, and on the other from Sweden's rocky mass, Belgium and France on the other. These two straits, intersecting one another, were producing mollusca on a prodigious scale, and these not only became in some unaccountable manner chalk, but piled themselves up thousands of feet deep in places, and in considerable heights, not, you observe, in any regular layer or layers, but in indiscriminate masses with big gaps between. ...
>
> " Why, if this were the origin of chalk, cannot our geologists point to some part of the ocean or seas where chalk is now under process of formation? ..."

We recall that geologists assert that "the present is the key to the past" (the Uniformity Principle), so Beaumont challenged them to make good on this claim. And then he went on to conclude with this thought.

> " In all these matters there is only one scientific explanation which covers the entire problem. The chalk was laid down, as the parallel ranges of mountains were laid down, by the deposit of materia from other worlds and on several occasions far separated in space and time. The mollusca forming the chalk were subjected to treatment that explains why chalk is found in big patches, or strips, and in other parts of the earth it does not exist at all. For if the shells of the *foraminifera* become chalk by gradual process of decay, chalk should be found in

The Great Flood

every part of the seas in small quantities. In the chalk deposits we have, on the contrary, the accumulated mass of all the mollusca in entire worlds, worlds perhaps, where they were commoner than in our seas, masses of chalk flattened out in the process of passage, sifted, bleached, carried along at tremendous speeds, subjected to tremendous magnetic strain, and finally dumped down in a huge mass, or masses under vast pressure. ..."

In the light of our present findings we would have to grant that his conclusion is essentially correct, except perhaps for the time scale, even though it cannot be derived plausibly from his conception of the nature of comets. However, his tracing of volcanism, earthquakes and extremes in the weather likewise to *meteoric* events was misguided. In fact, he was able to entertain these ideas only because of his hazy understanding of basic physics, especially orbital mechanics, and his reference to "... tremendous magnetic strain ..." also lacks foundation. So one can understand why his picture was poorly received by academicians of the time (and since). Nevertheless, his conclusions were much closer to the mark than those of orthodox science even today.

Continuing, now, with our reconstruction of that great catastrophe, if the fossiliferous rocks were mere side effects of that cometary encounter then much greater violence would have resulted from the impact of the palpable nucleus. Moreover, in view of our considerations on volcanism in Appendix D and the great number of volcanoes distributed around the Pacific Rim we already know the most likely place to look for that impact. Presumably the object penetrated the crust of the earth and transferred some of its momentum to the quasi-fluid interior, giving rise to a circulatory motion that quickly died away. In the absence of any other known mechanism one can easily conclude that this was the motion that gave rise to the separation of the continents. Those who are not already familiar with the fact of this separation may welcome a brief review of the case.

Men have long noted that the eastern borders of the Americas would fit well against the western borders of Europe and Africa. Indeed the fit is so remarkable, especially along the contour of the continental shelves, as to suggest that the Americas split away from what was once a very great primordial land mass, often dubbed "Gondwanaland". But not until fairly recently could that proposition be considered confirmed. The proof was found almost by chance in careful observations of the (slight) magnetization of the ocean floor. Magnetic surveys of the sea floors show discernible bands of magnetization running parallel to the mid-ocean ridges in which, between adjacent bands, the direction of magnetization changes. The tell-tale clue, then, is that the pattern of these bands *tends to be symmetrical* with respect to the central ridge [16; p.81].

The unavoidable conclusion is that the floor of the ocean has spread apart from the mid-ocean ridge allowing new rock in a fluid state to well up from below filling the gap. The magnetic impurities in this newly solidified rock tend to be aligned along the direction of the magnetic field that prevailed at the time, thus giving rise to that slight magnetization of the sea-floor. The alternating bands show that the magnetic field changed direction several times during the separation, and the symmetry of the bands with respect to the central ridge proves the separation.

According to prevailing theory the separation is deemed to be a slow, continuing process. The slightly fluid interior of the earth is thought to be driven into convective circulation by heat generated by radioactive decay in the depths, and the continents are thought to be carried along with the motion. According to this picture, which has come to be called plate tectonics, the crust of the earth has divided itself into a number of irregularly shaped "plates" that continually grow above regions of upward convection and vanish along others where, at opposing fronts, one of the plates is thought to dive cleanly

The Great Flood

under the other one and return to the depths.

But although one must admit plastic flow in response to extreme pressure in the very hot interior of the earth he is hard-pressed to imagine that cool, crystaline crustal rocks would behave in this fluid-like fashion as well. Manifestly they do not. However tracing the separation to a cometary collision not only provides for the impetus behind the motion, it also accounts for the magnetic anomalies in the ocean floor. Namely, in that case the residual magnetization in the sea floor does not record the direction of an equilibrium terrestrial dipole field, but that of the *transient* field that prevailed immediately after the impact. This field would have been more directly traceable to the decaying comet, and its configuration could easily have changed rapidly as explained in Appendix C. One might note that a rapid separation of this kind would have exposed vast areas of very hot rock to the ocean waters, giving rise to greatly increased evaporation into the atmosphere. Thus could one account for the forty days of rain that are cited in the Bible as the cause of that flood.

In addition to volcanism and seismic activity we must include violent atmospheric phenomena among the after-effects of that event. Even rain was unknown beforehand, for according to Chapter 8 of the Book of Genesis, verses 5 and 6:

" ... for the Lord God had not caused it to rain upon the earth, and there was not a man to till the ground.
" But there went up a mist from the earth and watered the whole face of the ground."

Of course, this is exactly what one would expect of an atmosphere obeying the normal laws of physics so the statement hardly needs corroboration, but it is interesting to note that similar conditions have been observed on occasion even in relatively modern times. In fact, quoting one elderly keeper of the ancient Indian lore Vine Deloria reports [27; p.234]:

" ... When the world was young, the land east of where the

Cascade Mountains now stand became very dry. This was in the early days before rains came to the earth. In the beginning of the world, moisture went up through the ground, but for some reason it stopped coming."

In this light it is easy to conclude that thunderstorms and the other extremes of weather are indeed bizarre aftermaths of past cometary encounters.

We can hardly leave this subject without addressing the question of when that great event took place. In view of the bizarre origin of the rocks dating techniques based on radiative decay rates are seen to be entirely irrelevant; they have no bearing on the problem whatever, and other standard lines of reasoning are similarly to be questioned. Calculating age from plausible rates of erosion and deposition is a remote possiblity, but normal erosion and deposition played only a minimal part in forming the earth. Thus, the age question appears to be an exceedingly difficult one, and this writer can suggest no satisfactory way to resolve it.

Surprisingly enough the date given in The Bible may well be correct. As is well known, this works out to about 2350 B.C. Countering this, on the other hand, we are told that the documented history of the Nile Valley extends back to about 3100 B.C. and leaves no room for any such flood. As it happens, the prevailing reconstruction of those olden times is based on the work of one Manetho, an Egyptian priest who wrote shortly after the Greek conquest—perhaps about 300 B.C. He it was who first collected the various kings into dynasties and assigned years to their reigns, though modern scholars have amended his imaginative chronology considerably. But problems with this modern system have already been pointed out. Velikovsky showed evidence that the Exodus had to be fitted into the Egyptian scheme near the end of the 13th dynasty—not during the 19th as presently conceived. He then identified the subsequent Second Intermediate Period as the era of collapse

The Great Flood

following the Exodus, which, as we now understand, was simply a local view of the world-wide eclipse following the cometary impact of about 1445 B.C.

Later on, Courville [24] proposed that Manetho had artificially lengthened the history of his people simply in order to impress the Greeks who then ruled over Egypt—this by lining up the various kings in sequence when in fact some of them had ruled concurrently. In particular, Courville was able to fit the surviving evidence into a system wherein the various kings at Memphis ruled at the same time as kings at Thebes, in upper Egypt. According to his reconstruction the Exodus took place during the 7th dynasty (as viewed from Memphis), making the First Intermediate Period actually the same span of time as the Second. Seen in this light the first dynasty would have begun at Thinis in the vicinity of 2100 B.C., a thousand years later than currently imagined and 250 years later than the time given for the flood instead of 750 years before it.

In the absence of a precedent one might hesitate to imagine that the world's Egyptologists, careful scientists all, could err so grievously in their reconstruction of the times, but we have already observed an error of similar proportions current among authorities on the ice ages. Here, over the course of more than a hundred years, these comparably astute scientists have traced the various advances and retreats of their proposed ice in detail without ever discovering that the ice, in fact, never existed. Surely, if one group of scientist can err so remarkably then others are in danger of doing so as well—and for the same reason, a reason that will be examined in a later context.

Remarkably enough, even the pre-Cambrian granites seem to have been tainted by that bizarre phenomenon so the customary radiometric dating methods have to be abandoned altogether. A brief discussion of this surprising fact will form an apt finale to our sojourn in antiquity so let us consider it now.

A word of explanation may be in order first of all. The so-called *pleochroic halos* are small, spherically symmetric colored regions observable in some natural crystals under a microscope. They result from radiation damage caused by alpha particles emitted from minute specks of radioactive inclusions. In the course of time the crystal lattice accumulates damage due to the passage of these particles, which damage shows up as a discoloration spherically disposed around the central grain. The coloration ends abruptly at a radius set by the initial energy of the particles, so when alpha-particles of different energies are emitted from that central point each group gives rise to its own halo. In cross section, these haloes show up as a series of concentric rings, and one can determine the radioactive species causing the rings from measurements of the various radii. In other words, a given radioactive species will generate its own characteristic set of halos.

In particular, a grain of Uranium^{238} gives rise in principle to eight concentric halos*, generated in turn by its own alpha decay, and seven more that result from the seven alpha-emitting isotopes in its decay chain. These are U^{234}, Thorium^{230}, Radium^{226}, Radon^{222}, Polonium^{218}, Polonium^{214} and Polonium^{210}. Now Robert Gentry's remarkable and surprising discovery [36] is that *halos arising from the polonium isotopes are often found alone—unaccompanied by those arising from its parents in the above chain.* Namely, he observed the set of three concentric halos arising from Po^{218} and its daughters Po^{214} and Po^{210}. Likewise he saw the pair originating from Po^{214} and Po^{210}, as also single halos stemming from Po^{210} alone. The profound riddle in all this stems from the very short half lives of these three radioisotopes. They are, respectively, 3 minutes, 164 microseconds, and 138 days. Now then, given these very short half-lives and the prevailing views about the origin of the earth, where did these inclusions of pure polonium come from?

---

* *Some are so nearly the same radius that they cannot all be resolved.*

The Great Flood

Because of the short lifetimes of the polonium isotopes and their intermediaries in the decay chain no native polonium can exist anywhere today; every last bit of it must have derived very recently from Radon^{222} in keeping with the decay scheme shown above*. Did the granite then stem from a time in the remote past before all the native polonium had decayed away? Even that won't do, for if the granite had been formed from a melt, as geologists assert, then any isolated polonium that might have existed in the melt would already have decayed away before the mica could have crystalized—especially so the Po^{218} with its 3-minute half-life since it has no immediate parent other than the Radon^{222}. Gentry could conceive of only one resolution to the riddle; he proposed that the granite had simply been created by God during that first day, and some specks of polonium just happened to be created along with it. He therefore offered the polonium halos as proof that the earth was created out of nothing and did not condense from primordial dust as modern cosmologists insist.

Without a doubt the polonium halos do disprove the Principle of Uniformity, but they fall short of proving the six-day creation since, in the light of our newfound understanding of nature, two alternative interpretations can be given for them. Namely, we might conclude that fully formed granite

---

* *Let it be noted that the picture is slightly more complex than described above because the two beta-emitters, Lead^{214} and Bismuth^{214} separate Po^{218} and Po^{214}. These have half-lives of 27 minutes and 20 minutes respectively. Likewise Lead^{210} and Bismuth^{210} separate Po^{214} and Po^{210}. They have half-lives for beta decay of 22 years and 5 days, respectively. All of these times are, of course, infinitesimally small compared to the imagined geological time scale. One might note in passing that a typical Po^{210} halo, for example, would be indistinguishable from an inclusion of either Lead^{210} or Bismuth^{210} because the beta particles (electrons) emitted by the lead and bismuth, being so very much lighter than the alpha particles, cause no visible damage to the crystal.*

emerged from out of that fourth dimension, Fortean fashion, following that great cometary impact. During the process it might have picked up inclusions of polonium in much the same way that those snails in the loessian nodules picked up the limey silt inside their bodies. Alternatively, one might imagine that a fine dust consisting of all manner of materials emerged widely in our world—again in Fortean fashion as a result of the comet. Here, too, some of that dust could have materialized within pre-existing materials, and if that dust had contained small grains of polonium then that which formed within the mica would have proceeded to generate halos. This would seem to be the easiest explanation for the halos even though it offers no accounting for the ultimate source of the polonium. However, in a regime where elements may be transmuted and palpable material can materialize out of nothing at all, perhaps this a point that we need not labor over at length.

As one stands back and views that grim holocaust through his mind's eye he can only shrink in horror, for even the worst that can be imagined must fall short of the reality. The earth breaking apart, the unleashing of that incredible amount of energy and the twisting of worlds on top of each other stretch even a vivid imagination to the breaking point. That the elite doctors of science can gaze out upon the wreckage and see naught but age after age of tranquility must be counted among the wonders of our time—one that merits attention for its own sake. We will consider it briefly in the pages to follow.

Epilogue:

AFTERTHOUGHTS

THE GREAT CATACLYSM that laid waste to much of the earth sometime before the dawn of history has now claimed one more victim; having come forth to be recognized it has destroyed all of modern geology as well. Of course, those who were alive to the evidence have understood for many years that Uniformitarian geology was already dead, but it refused to lie down in a seemly fashion because no other interpretation for the earth's surface could be offered in its place. But now that theory can now be laid to rest at last, and in an adjoining grave can be buried also the Theory of Evolution since it is moribund without Uniformitarian geology for support. Now then, as is customary on such occasions, we might pause for a moment to say a few words in memory of the deceased.

The trouble all started, we recall, with that revolutionary movement in Eighteeenth-century Europe that has come to be known as The Enlightenment. It was spawned by an intelligentsia who chafed under constraints and dogmas imposed on society at large by the Church. In fact, those elite thinkers traced many ills of the day to the workings of a clergy who were badly out of touch with reality.

In keeping with their rationalistic view of nature those free-thinking revolutionaries set aside any thought that a supernatural agency might have formed the earth, urging instead that our planet developed through natural processes operating in accord with the customary laws of physics and chemistry —even as we see them in operation today. Though it may be variously stated this Principle of Uniformity in effect defined the ground rules to which they thought that all of science should conform. Its intended purpose was to deny the Hand of God as an earth-shaping force, but it did a great deal more than that as we have now discovered. In fact it rendered impossible any valid understanding of earth's history, to say nothing of the cosmos at large.

One must marvel that those people should have thought to impose rules upon Nature herself—rules conceived in their own heads without even a pretense of foundational evidence. Also remarkable is the fact that they began as a small minority and yet were able to grow into a ruling majority in academia within the space of a few years even though their picture of nature was so vastly in error. This unlikely flourishing of so obvious a mistake constitutes something of a marvel in its own right and deserves our attention briefly.

As we know, Enlightened thinkers hold to a strictly materialistic view of man. They look upon him as a mere bag of chemicals all interacting in accord with the immutable laws of nature. Certainly there is no thought here of a transcendental aspect of life or of a Creator. Indeed, they trace the origin of all life to a kind of primordial mother soup, out of which it evolved in stages over great aeons of time. It follows that they would also have to deny any absolute standard of right and wrong required of all men to uphold. Thus, Enlightened man can have at best only a rudimentary sense of morals and ethics, and he probably has but a vague sense of personal honor also. This is not to say that he is insensitive to the esteem of his fellows, but how he gains that esteem—whether honestly or by secret deceits would presumably make little difference to him. Outer appearances are all that can matter because he has no concern

Afterthoughts

for a transcendental soul that might be marred by chicanery —by stains that might one day stand out for all to see. The only rein on his behavior would be fear that his frauds might be discovered in the here and now.

Even a blind man should be able to see that a contest between individuals holding these views and those constrained by the customary standards of deportment would be no contest whatever, and this "tilting of the playing field" would be as decisive in academia as anywhere else. Dishonest scholarship can be very compelling indeed, and we have seen many examples in these pages where sound evidence has been deliberately misinterpreted or blandly ignored for partisan advantage. Thus it came to pass that the Enlightened class became dominant in most institutions of higher learning, and the Principle of Uniformity came to have the force of law in the sciences *even though utterly invalid.*

We saw the irresistible pressure of this law in operation already more than a hundred years ago when Skertchly and Kingsmill reported on those great limey slabs lying upon the loess in China. Although they probably described the slabs accurately one can scarcely imagine that they truly believed the explanation that they offered for them. Could they, for example, really have thought that old rivers would have flowed upon the surface of the loess instead of wearing through to the basement rock? Or that unlike river beds today the old ones would cement together to form a conglomerate? Or that such river beds would stand out as bosses and banks upon the loess plain? Or that fragments of old river beds 6 or 7 feet wide and 20 to 30 feet long would project like shelves from the loess? Or that five or six residues of old rivers would cling to the valley sides at different levels? Or that 80 percent of their courses would be independent of the slope of the land? Or that the beds would cross each other?

We ought to grant them more sense than that. Since they went to the trouble of enumerating so many puzzling features of those "old river gravels" one has to conclude that they wanted their readers to know that they were offering that foolish interpretation under duress. Now the reader should be aware that any paper submitted to a professional journal is referred by the Editor to some recognized expert in that same field for evaluation. If this established expert deems a paper unsuitable then the journal will not publish it. The system is defended on grounds that it tends to preserve high standards and avoid duplication, but it also helps to preserve prevailing errors in place. Namely, a referee with little sense of propriety might reject a prospective paper for the simple reason that it would threaten his own dignity, status or claim to expertise. That is to say he might attempt to preserve the *status quo* out of narrow self-interest, whatever its follies.

In keeping with this system of referees Skertchly and Kingsmill were probably told flatly that their paper could not be published unless they interpreted the slabs in a manner agreeable to the Uniformity Principle. As is well known, having papers published in the professional journals is very important for those with academic aspirations. It leads to security of position, and without a growing list of published papers one hardly has a future in the academic world. Certainly the referee was at fault for insisting that truth be sacrificed in order to defend the Uniformity Principle, but the authors are also to be blamed since by consenting to "ride with the tide" they became part of the tide themselves.

But Skertchly and Kingsmill's trivializing of their evidence should have fooled no one. Usually such revealing facts are either not discussed at all or they are rendered void with more finesse. For example, in his impressive compendium on Glacial and Quaternary Geology, *wherein the bibliography ran to 65 pages of references in ten languages,* Flint [33] devoted a mere 20 pages to the loess out of some 800 pages of text. He described some of its properties and its distribution around the world, but he made no mention whatever of Winchell's highly

Afterthoughts

significant report about the loess grading smoothly into the drift. And neither did he mention Skertchly and Kingsmill's observations about those great limey slabs lying upon the loess in China. Nor did he note those occasional pebbles in the loess or the points made by Berg, Howorth or Keilhack. And most significantly of all he made no mention whatever of those limey nodules whose internal structure we found so revealing. However one might try to be fair he can hardly imagine that these omissions were the result of accidental oversight. On the contrary, their common import suggests that the author deliberately suppressed those remarkable properties of the loess in order to conceal an obvious failure of the Uniformity Principle and thereby render his treatise doctrinally acceptable.

Other examples of this sacrifice of truth for ulterior purpose are not hard to find. As another case in point we might note how the authors of one recent text in elementary geology represented the Uniformity Principle [37;p.18]:

" The Uniformitarian Principle, like any other scientific generalization, rests on the circumstance that no known facts contradict it; all can be interpreted in accordance with presently operating physical, chemical, and biological processes. Geologic study through generations has failed to find evidence of ancient processes totally unlike those existing today."

The words speak for themselves. If they had been written two hundred years ago one might excuse them as the product of mere ignorance, but having been written in modern times, with more complete data at hand, they betray the authors' willing lack of candor and give the appearance more of indoctrination than instruction.

Nor have geologists been alone in overlooking evidence that would put The Enlightenment in bad repute; some biologists are equally culpable. In fact, one of the most revealing phenomena ever to come to my attention is well known amongst psychologists, though one seldom hears it discussed in public. I have in mind the so-called "autistic savant syndrome", the name given to a peculiar condition in which some severely retarded persons display extraordinary talent in a specific narrow field. Such people might show inexplicable skill at mental arithmetic, for example, or in art or music.

The case of Leslie Lemke offers a striking example of this condition [62]. Lemke was blind from birth, spastic to the point where he could neither speak nor walk unaided, and he was severely retarded as well. Nevertheless, at the age of 16, in the middle of one night, he crawled up onto a piano stool and began to play—not haltingly, but as an skilled musician. From that day forth he had only to hear a piece of music once and he could play it back in a polished manner on the very first try. On that first occasion he played Tchaikovsky's First Piano Concerto, which was familiar to him as the theme music of a popular entertainer.

Now being blind Lemke has never seen a keyboard, and being spastic he has only marginal control over his fingers in other circumstances. In particular, he is unable to walk without support, and at the table he cannot even manage his silverware —yet seated at the piano he is a flawless performer. Quite obviously, any attempt to accommodate this phenomenon to the Enlightened view of man would be, in the extreme, ridiculous. But perhaps one can vaguely perceive the mechanism behind the phenomenon even if he cannot understand exactly how it works. Namely, we have only to note that other dimension of space and grant him an extension along it. Then on those musical occasions control of the muscles in Lemke's arms and fingers is taken over by some other "personage" who resides along that other dimension and who manipulates those muscles in the manner of a puppeteer* working a marionette.

---

* *One small problem with this interpretation is that Lemke obviously enjoys the experience and certainly appears to be the active force behind the playing rather than merely a passive bystander. However,*

Afterthoughts

Let us note that one is not obliged *to assume* that extra dimension into existence specifically to formulate this picture; its existence has already been proved from other evidence entirely. He has only to note in passing that this added extension of reality provides a basis for describing the phenomenon. One might go further and conclude that all living organisms have an extension along this extra dimension; presumably without it they would revert to the mere material and die*. In any case, we seem to have little choice but to identify this extension of the human being as that very "soul" which is so stoutly denied

---

one need not assume that the effect on him is necessarily limited to the nerves which activate his muscles; he could easily receive other sensations and impressions as well. Furthermore, it is entirely unnecessary for our present purpose to inquire into the identity of that 'puppet-master'. Some have urged that the phenomenon must be traced to the direct intervention of God Himself, but arguing against this view, in my opinion, is the fact that Lemke is in no sense healed of his affliction. When not seated at the piano he is as helplessly spastic as ever. One might ask why this hypothetical puppeteer restricts himself to merely playing the piano and will not assist Lemke in other ways—help him to handle his silverware, for example. He is certainly able to help. On the night of his first playing, for example, he helped Lemke out of bed and up onto the piano stool—feats which he was quite unable to perform on his own. I conclude that this agent has been given specific instructions about what he is to do and what he is not to do. As of this writing Lemke is 42 years of age and lives on a farm in Wisconsin with his step-sister.

* *In my view the mystery of life begins at the cellular level. Here individual molecules are <u>as if drawn together by specific long-range forces</u> so that combinations take place very much more rapidly than the principles of thermodynamics would allow. This bee-hive of directed activity is truly beyond understanding. However, recognizing the cell as a three-dimensional cross-section of a four-dimensional entity one can at least understand why. The theory that life sprang spontaneously from some primordial pond seems all the more implausible in this light.*

by the Enlightened intelligentsia*.

Some readers might quarrel with a word here or a word there in my interpretation of the phenomenon of Leslie Lemke, but there can be no quarrel about its general import. It clearly proves the transcendental essence of the human being and thereby just as surely *disproves* the Enlightened view of man. Here, then, we see another lapse from the touted standard of science. Honest seekers after truth would have shouted this phenomenon far and wide; learned papers would have been written explaining its obvious implications, but nothing of the sort has happened; such cases are ignored, and the Enlightened view of man remains unchallenged in academic circles.

Let us take a few moments to review some of the other instances that we have seen already where scientists have either falsified evidence or ignored its plain implications. Firstly, of course, were the loessian nodules. Science responds by casting a pall over these objects and ignoring them as much as possible; not a word is ever written about their structure. And then we recall those great limey slabs lying upon the loess in China. Science explained them away with a grotesque absurdity and then let them lay in peace ever after. In Appen-

---

* *In Chapter 8 we saw that comets, apart from their condensed nucleus, are objects in four dimensions and manifest little or no ponderable mass. In Appendix D we deduce that material in this extended state does not resolve into atoms in the normal fashion so we dubbed it "hypermatter". Since the soul displays no ponderable mass then it might be similarly composed of hypermatter. In comets this substance is unstable; it condenses back into normal matter during the course of time. However as an extension of a living organism it appears to be more stable. Or, if it does condense into ponderable matter after the body's death then it must do so in some less-than-obvious way. As one possibility it might simply condense into the atmosphere around and about. In a closed system this mode of condensation ought to be observable as a change in air pressure and perhaps also of composition.*

Afterthoughts

dix A we found that Laboratory physics does not describe the lunar tides accurately; the theory has to be "fudged" in order to fit, but this fact is never acknowledged in print. Then we found that the scientific theory of comets similarly does not square with their observed properties, but the conflicts are ignored, and the theory is retained anyway. Likewise the polonium haloes in granitic mica are a definitive counter-example to the Uniformity Principle, but the haloes are ignored and the Principle is retained regardless. Even the common thunderhead and the winds of passion refuse to accord with Laboratory science, but this fact also is ignored. And the list evidently goes on because William James probably had none of these cases in mind when he wrote, "around and about the accredited and orderly facts of every science there ever floats a sort of dust-cloud of exceptional observances—an unresolved residuum of occurrences, minute and irregular, which always prove easier to ignore than to address forthrightly."

As we know, scientists like to boast about the vast difference between science and religion. Religions, they suggest, derive from childish myths which are preserved by those of simple mind who cannot face up to reality, whereas they represent science as the hard-headed study of Nature as she actually is. But we now see that this is a mere pretense. For Science is instead the attempt to "squeeze" nature into an artificially simplified Laboratory space—one conformable to human reason, and those attributes that refuse to fit into this Laboratory are conveniently ignored. Thus, despite its grand pretentions, Science has become the *menial servant of a demonstrably false religion*—one that deifies human wisdom, such as it it is. It is surely ironic that after having developed in reaction to a priesthood supposedly out of touch with reality the Enlightenment should have spawned a priesthood of its own who are genuinely out of touch with reality—a priesthood, indeed, who have foisted upon a credulous public arguably the most pernicious imposture that the world has ever known.

I wish it were possible to excuse this deception as a mere blunder—one that arose more from carelessness than design. However, even in this short study we have seen many occasions where an honest appraisal of the data would have put the matter right. But it was not put right. Insidiously, the illusion has a life of its own and proliferates despite the evidences that can be arrayed against it. I stress the point because it has overflowed the ivied halls and proclaiming the Authority of Science threatens to undermine the very foundations of civilized society. To bring the problem into perspective let us observe how one group of activists are promoting it today and note what they hope to accomplish with it. These people call their faith Humanism, but it remains essentially the same doctrine as before. Here are its first three tenets [54;p.8]:

" First: Religious humanists regard the universe as self-existing and not created.
" Second: Humanism believes that man is a part of nature and that he has emerged as a result of a continuing process.
" Third: Holding an organic view of life, humanists find that the traditional dualism of mind and body must be rejected."

Thus, they endorse the Enlightened view of man, denying him any transcendental component whatever. They continue by noting a few plain implications of this viewpoint so we are not obliged to figure them out for ourselves. Accordingly, another tenet listed in this *Humanist Manifesto* is:

" *Fifth:* Humanism asserts that the nature of the universe depicted by modern science makes unacceptable any supernatural or cosmic guarantees of human values. Obviously, humanism does not deny the possiblity of realities as yet undiscovered, but it does insist that the way to determine the existence and value of any and all reality is by means of intelligent inquiry and by the assessment of their relation to human needs. Religion must formulate its hopes and plans in the light of scientific spirit and method."

Afterthoughts

What we see here might be compared to a snake that has turned full circle and bitten its own tail. For there can be no doubt that the compilers of this document were thoroughly convinced that in Science they had found Ultimate Truth. So certain were they, indeed, that they felt free to pronounce in peremptory tones how the benighted needed to adjust their way of thinking. The terrible irony in all this is that those Enlightened zealots were doubtless unaware that they had been duped by others of like mind who had falsely reported the testimony of Nature—partly to promote that very doctrine and partly to secure places for themselves in the academic establishment. Skertchly and Kingsmill might be offered as examples of such.

Now let us consider a few more tenets of that movement and find out why it poses a threat to society at large. The fifth tenet given above is as close as that document comes to touching on morals and ethics, but a second version, *The Humanist Manifesto II*, issued in 1978, addresses this vital topic explicitly in these words [p.17] (emphasis as in the original):

" *Third:* We affirm that moral values derive their source from human experience. Ethics is *autonomous* and *situational,* needing no theological or ideological sanction. Ethics stems from human needs and interest ..."

Here is a frank statement of a conclusion that we had already reached for ourselves; Enlightened man recognizes no "binding" standard of ethics whatever. In effect, he is free to define his own standard in keeping with his own personal ambitions. We have already seen how this attitude played out in one small segment of society—namely, science departments in the major universities. There, by willfully misrepresenting the evidence, or ignoring it altogether, Enlightened partisans were able to impose upon the academic community a grotesquely false view of earth's history. But this was only one small corner of society. As this doctrine continues to propagate more and more people will accept the idea that their own advancement serves some "higher good" and therefore will excuse themselves from the customary standards of behavior.

Indeed, I suspect that any Enlightened individual who pilfers from the the commonwealth is able to adjust his conscience in such a way as to render the practice acceptable. Probably the framers of that document would assure us that this is not what they mean by situational ethics, but their protestations would hardly matter because the seeds of knavery reside within the very tenets of the Enlightenment. Since Enlightened man has no soul to be stained by his malpractice, and since he recognizes no ultimate standard of right and wrong, the only force that can restrain his looting of the commonwealth is the strong arm of the law close at hand*. But even then clever individuals who feel no restraint of conscience will lie awake nights devising ways to circumvent the law so they can pilfer in safety. This is the frame of mind that must abound in an Enlightened society, and although the trail is not easy to follow I believe this mentality at work accounts for the widespread poverty that we see today in the midst of plenty. Indeed, if the present trend continues, and the national ethic continues to erode, I anticipate that the country must ultimately revert to a police state or go bankrupt.

By contrast I like to think that every stable and prosperous society maintains its vigor only so long as it frankly acknowledged the *spiritual nature and destiny of man.* Namely, the people in successful societies yield to a sense of personal

---

* *I think this is a fair statement when understood in a general sense even though it may be utterly inappropriate for specific individuals. The dichotomy is indistinct partly because one's behavior is governed not by reason alone but by the emotions also, and they can be compelling at times. And furthermore, some convinced rationalists are constrained by conscience in spite of themselves. Indeed, I myself have known professing humanists who were paragons of virtue, and of course soul-filled rogues are common as well.*

Afterthoughts

honor and *govern their own behavior within narrow limits* in order to "please the gods" and to avoid staining the timeless essence that abides within us all. A society composed of such individuals would need few policemen to maintain order and preserve the commonwealth intact.

As added points of interest let us note in passing a few more articles of faith from the Humanist Manifesto II (emphasis as in the original):

" *Sixth:* In the area of sexuality, we believe that intolerant attitudes, often cultivated by orthodox religions and puritanical cultures, unduly repress sexual conduct. The right to birth control, abortion and divorce should be recognized. While we do not approve of exploitive, denigrating forms of sexual expression, neither do we wish to prohibit, by law or social sanction, sexual behavior between consenting adults. ..."

" *Eleventh: The principle of moral equality* must be furthered through the elimination of all discrimination based on race, religion, sex, age or national origin. This means equality of opportunity and recognition of talent and merit. Individuals should be encouraged to contribute to their own betterment. If unable then society should provide the means to satisfy their basic economic, health and cultural needs. ..."

" *Twelfth:* We deplore the division of human kind on nationalistic grounds. We have reached a turning point in human history where the best option is to *transcend the limits of national sovereignty* and move toward the building of a world community in which all sectors of the human family can participate. ..."

In short, the Humanist creed and agenda define what has come to be considered "Politically Correct" in today's world; they are drummed into us daily in the communications media, and they are taught in the public schools at every level. Humanism is therefore a force to be reckoned with. I tremble at the thought that these activists would remake the social order by the light of that false view of man and perverse sense of morals and ethics. I am especially dubious of those who would exercise power in that grand social order. For want of any spiritual perception of self-worth those Enlightened elite must rate themselves chiefly in proportion to their material wealth so I would anticipate no limit to their greed. Since they would also lack the customary ethical constraints, neither would I expect them to limit the means by which they might gratify it. Thus, despite the Humanists' egalitarian vision I would expect an ever-widening gulf between those wielding power and the masses who would be obliged to support them. As their doctrine continues to spread that excessive stratification of wealth must propagate into every segment of society.

Wolfgang von Goethe, the noted German writer, once observed that he could think of nothing more frightening than ignorance in motion. In retrospect we can be certain that he never paused to contemplate *delusion* in motion—especially delusion so fetchingly clothed as to seduce even the intellectual elite. As we now understand, the so-called Enlightenment is in truth a sorry delusion, one that has wrought untold strife down through the centuries, and it continues to work mischief still today. Its cost to society includes countless lives and vast fortunes spent trying to squeeze the deep mysteries of nature into that simplistic scientific Laboratory. But such costs pale in comparison to others that may not be so obvious. Thus, by propagating this imposture the doctors in academia blind their students to the more meaningful aspects of life and thereby condemn them to a world of follies. Some may come to a better understanding eventually, but many go to their graves thinking themselves wise in their Enlightenment—easily the greatest folly of them all.

However, much more important to society at large is the fact that, by equating themselves to a bag of chemicals, converts to this doctrine abandon any sense of honor, truth, virtue, duty, loyalty, sobriety, morals or ethics. These are the cares of man mindful of his immortal soul; they are of no concern to a

Afterthoughts

bag of chemicals. Consequently, by propagating their perverse doctrine those doctors send into the world a class of predators willing to prey upon their neighbors and upon society at large without scruple*. Because of its obvious threat to the general welfare their doctrine would be objectionable even if valid, but being palpably invalid it is all the more objectionable. Tragically, but perhaps also fortunately, most of those propagating the imposture are innocent victims of it themselves; having had no part in falsifying Nature's testimony they merely took the scientists of former times at their word. Therefore I dare to hope that they, at least, in the light of better understanding, might be moved to renounce that pernicious doctrine and establish themselves on some other foundation. I feel confident that by setting aside this delusion our society will eventually recover from the ills that have befallen it of late.

Perhaps the reader will permit me a brief lapse into the frankly problematical for one final thought that may be of interest to some. Namely, we saw in Chapter 12 that the earth did indeed suffer a great catastrophe that answers well to the Deluge described in the Book of Genesis—excepting only that I traced it to the impact of a comet instead of the spoken Word of God. But even if it were the result of natural forces in operation God is said to have known about it more than a hundred years in advance, giving Noah time to build the ark. This would imply a supernatural foreknowledge, not only of the event itself, but also that it was going to be of catastrophic import. Likewise the destruction of Sodom and Gomorrah was known in advance, thereby allowing Lot and his family a chance to escape the coming destruction. Each of these events is said to have been a judgement of God upon wicked people.

Unfortunately I perceive that this predatory attitude prevails widely today—a fact that I trace in large measure to the Humanistic indoctrination of children in the public schools. However, I gladly acknowledge that individual exceptions to this trend are not uncommon.

Should we understand, then, that the great holocaust at Cibola was similarly earned as the wages of sin? Had the Cities become so depraved during the years since Fray Marcos that not a soul was allowed to survive that he might sing their praises and mark their passing?

For my own part I tend to believe that Cibola and her neighbors succumbed to a simple accident of nature and that God bore them no ill will whatever—for reasons that may not be altogether persuasive, but I will share this view as a closing thought so the reader can judge the matter for himself. To this end let us return to Easter Island and contemplate the riddle of those huge statues one more time.

With only rare exceptions those who have offered to explain their mode of transport took for granted that, contrary to all indications, cordage and timbers were available at the time so there is really no problem after all. The people merely rigged up levers, rollers, and derricks so as to handle the loads in perfectly ordinary ways. But in simple fact the ancient folk had neither cordage nor timbers at hand. Moreover, one searches the island in vain for scars that would have been left by engineering emplacements or ramps or smoothed roads. And if more proof were needed it can be found in the boat houses that provided the best shelter for the people in those days. They testify plainly that the islanders did not have the essentials for heavy construction at their disposal. The fact that the king and his wives and chief priests all lived in the same rude huts as did the rank-and-file farm hands should have made this obvious from the beginning. Even if they had only primitive architectural vision they could still feel the cramp of close quarters and the chill wet winds of winter blowing through the grass walls of their hovels and yearn for relief. And the materials for relief were easily available; the king had only to bid the carvers to make blocks for walls and rafters to support a roof. One must conclude that he did not do so simply because the rafters once made could not be moved.

Afterthoughts

In fact, nowhere on the island is any stone or structure to be found, apart from the statues, their top-knots and mounting platforms, to suggest that the people were able to move heavy loads in those days. The unavoidable conclusion, then, is that the great statues moved by the power of mana*, just as the natives have explained from the beginning. Moreover, there was an intelligence with a will of its own behind this power. It would move statues of any size, but it would not move building blocks even for the king.

This suggests that it had a purpose of its own in encouraging the production of the statues and in arranging them around the island as it did. In that case every detail of those images may be of significance, but for the present let us note merely that while they were made in a great range of sizes they were very similar in other respects—even including their facial expression. Perhaps this indicates that they were meant to be a group with a common spirit and not merely a random crowd. This idea is strengthened by the fact they they all faced inland; so they were a circle—a group with a common intent.

Now let us consider again that great social turmoil at the

---

* *Instances spring to mind of other prodigious works in stone that might have been inspired by similar unworldly intrusions into the affairs of men. The ancient ruin of Tiahuanaco, considered in Chapter 11, offers an apt case in point. Concerning its construction Hancock cites Indian legends, as passed along to early Spanish explorers, saying that the huge stones floated in the air, driven along by the sound of a trumpet [41;p.72]. The great pyramids in Egypt are other obvious examples, and the temple of Jupiter-Baal in Baalbek, Lebanon is still another. According to the Encyclopedia Britannica this remarkable temple contains the three largest stones ever used in construction. They measure more than 13 feet in height and breadth and are over 63 feet long! The only possible reason for hewing such huge stones was <u>precisely because they would not have to be moved by manual labor</u>, and the thrill of watching them move in this unearthly manner would have been all the greater the larger the stone.*

time of Prince Rokoroko He Tau which Métraux mentioned. We recall that the people refused to carry before the rightful King the standards appropriate to the royal dignity. Clearly this was not because the young prince had such great manic power as Métraux suggested, for in fact he had none at all. The reason can only have been the failure of the King's own power—or at least part of it. Obviously the statues quit moving at some point, and this must have been the time. Presumably this is why work stopped at the quarry; there would be no point in carving on the statues if they would not move when finished. And this also accounts for the importance attached to those newly fallen stones; perhaps the Stout People thought that they were somehow poisoning the mana and that the power would return if the island could be rid of them altogether.

If the alien stones fell at the time of the comet then the statues quit moving upon that occasion as well; apparently they were ready to play their assigned role. For a clue to what that role might have been we might note the expression upon their faces. Most obvious is the fact that their lips protrude—the universal appearance of one holding back the tears. But is it universal? As a check I asked some of the Easter Island natives what emotion they read upon the faces of the statues. Without revealing my purpose I inquired of six as opportunity permitted. Of these, two answered "deep thought"; two more "sadness"; another "sorrow", and the sixth "anguish". So I am not alone in perceiving that the statues were expressing *grief.* In that case their role was a poignant one. For the space of a few years after the comet the whole throng of them stood around in a circle grieving together as one—and then they departed the scene, their role finished.

I often ponder that sad tableau, and alone in the quiet of an evening I can imagine that those worthy Cities of Cibola did not die unmourned after all.

FINIS

Appendix A:

THE TIDES

ALTHOUGH ONE SELDOM hears mention of it the tides pose a profoundly disturbing riddle to prevailing science. Nearly everyone recognizes in a general way that the sun and the moon give rise to the tides, but in order to appreciate the riddle and its resolution we must go one step beyond the mere generalities and become familiar with the details. However, even before addressing the problem it will be necessary to make several brief excursions into other fields of study in order to acquire the key to understanding.

The first of these will be scarcely more than a pause along the way. Here we have only to note the recent discovery that spiral galaxies do not contain enough visible mass to account for their behavior under the action of gravity. It is as if our moon, keeping to its same orbit, were to complete its revolution about the earth in, say, ten days instead of twenty-seven. The earth's mass—namely its gravitating strength is simply not sufficient to sustain such a motion. And yet galaxies do behave strangely in this respect, at least those studied by Burstein and Rubin [20]. These observers measured the redshift in component stars, from which they deduced rotational velocity at various distances from the galactic center, and then they estimated the distribution of mass by counting the stars.

Surprisingly enough, they found that the rotation is not stable under the gravitational effect of the visible stars alone. In fact, in the cases they studied, *as much as 90% of the dynamically effective mass was invisible.*

But now that we recognize an added dimension of space it is only a step to conclude that part of this hidden mass may reside in other stars arrayed along that other dimension and that their gravitating effect is still felt in our plane of existence even though they may not be visible to the eye. This resolution to the riddle may seem too easy so let us weigh additional evidence showing that we experience gravitational effects from "alien" bodies disposed along that other dimension.

In particular, we might note that the gravitating mass of the earth itself is substantially greater than can be accounted for as rocky assemblage alone. Typical surface rocks, granite for example, have a density about 2.7 times that of water, and yet the earth's average apparent density is about twice that amount. Even when one recognizes a nominal increase in density in the depths owing to compression there still remains a huge excess in mass to be accounted for. Of course, as is well known, this excess is ordinarily attributed to a supposed dense metallic core at the center of the earth.

But a serious problem arises when one tries to correlate the observed rate of precession of the equinoxes—the "wobble" of the earth's axis of rotation—with this presumed distribution of mass*. This wobble is driven by the attraction of the sun and the moon on the equatorial bulge of the spinning earth, and it is strictly calculable given a particular distribution of mass. To get a feeling for the problem one might note that if he calculates simply on the basis of a uniformly dense interior for the

---

* *Unfortunately a fair statement of this problem relies on concepts which may not be familiar to the general reader. The writer begs the brief indulgence of those who feel themselves inadequately grounded in these matters.*

The Tides

earth he obtains a precession rate that agrees almost exactly with the observed value, and the agreement can be made exact by taking into account an increase in density toward the center that would result from increasing pressure. However, as already noted, the mass of that agreeably precessing body turns out to be scarcely more than half of the earth's observed mass as a gravitating body.

Efforts to make up this deficiency, then, must contrive a density distribution within the earth which yields the correct total mass and the observed rate of precession at the same time. Thereby arises the need for a very dense core, and it calls for a "balancing act". Namely, because of its great density the presumed core would have only a slight equatorial bulge and would therefore tend to precess very slowly. On the other hand the mantle above with its more pronounced bulge tends to a faster rate. In principle, then, by selecting densities suitably, these conflicting tendencies can be balanced in such a way as to achieve both the observed precession rate and correct mass simultaneously.

But such reckonings assume the core, crust and mantle to be rigidly connected so they can precess as a unit. However at the temperature prevailing near the center of the earth that dense core would likely be in a liquid state—and in any case the *outer core is certainly in a liquid state*, as is well known from the study of seismic waves. Therefore, that supposed dense central core would not be mechanically coupled to the mantle above, and if it were in a liquid state then neither would it respond as a rigid body to those torques imposed by the moon and the sun but would flow in response to them. The resultant motion of such a complex system is not easy to anticipate, but most probably it would not conform to the observed perfectly regular behavior of a nearly uniform spheroid.

Here again the obvious resolution to this seeming impasse is to conclude that this extra gravitating mass resides in other worlds which overlap the earth along that other dimension. However, this is not to suppose that we feel a gravitational effect of mass on those worlds equal to that of an identical mass residing here in our own plane. In fact, one can cite careful experiments which tend to confirm the existence of this overlapping mass and also show that its effects are indeed substantially attenuated.

That series of experiments was designed to measure the *gradient* of the earth's gravitational field below the surface. It was inspired by a suggestion that the force of gravity does not obey a simple inverse square law after all but includes a short-range component that vanishes for distances of astronomical significance. As a test of this conjecture Thomas and Vogel performed the said measurements at the Nevada Test Site [78;p.1173] where deep holes had been drilled into the desert floor for the purpose of testing atomic weapons underground.

These experimenters lowered sensitive measuring instruments into five such holes to depths of some 600 meters and noted the variation in gravity as a function of depth. It should be stated that the instruments are themselves exquisitely sensitive and very reliable. Ambiguities in interpreting the data stem from vagaries in the composition and uniformity of the deeply lying rocky substrate, not in the measurements themselves. But any such residual problems seem to have been very small because the gravity profiles measured in those five separated shafts agreed very closely amongst themselves and they all implied that the gravitating density of the rocky layer pierced by those holes was some 4% *larger* than their density as measured directly*.

---

* *One might note that, according to this picture, if the holes had been sufficiently deep the apparent density of the rocks pierced in this way would eventually become nearly twice the palpable density. That these authors observed only a 4% increase should be interpreted to mean that those other worlds are not all the same size. Perhaps only one of them was large enough to extend out to the altitude of the Nevada Test Site.*

The Tides

If one assumes that this excess in gravitational density derives from a mantle of extraterrestrial material overlying the local terrain then it would follow that the substance manifested only about 10% of the density of water. Since we know of nothing in nature with so low a density presumably the gravitational effect of mass in those other worlds is attenuated somewhat in their effect upon us. That is, their mass does not manifest here as strongly as would an identical mass here in our own plane. Since that excess apparent mass of the earth amounts to almost half of its total gravitating mass there could be a considerable number of other worlds of this kind.

Looking back on the developing solar system with these findings in mind one can imagine how primordial bodies which were suitably situated with respect to one another would, in the course of time, fall together and coalesce. Those that lay in the same world-plane would agglomerate in the normal sense, but those which lay in different planes along that other dimension would agglomerate "virtually" instead. That is, the energy of residual relative motion would be dissipated in non-elastic "collisions" until the various bodies would eventually come to relative rest overlapping one another even though they would not be physically touching.

Now then, with this understanding in hand let us return to our original topic and prepare to address the problem posed by the tides. To this end we might profitably review the actual calculation of two basic physical quantities, namely, gravitational attraction and centrifugal force.

Since the discovery by Isaac Newton men have understood that any material body exerts an attractive force on every other body—of strength given by

$$F = \frac{GmM}{r^2}$$

where m and M are the masses of the two bodies, r is their separation and G is a constant which depends on the units that we use to make the measurements. This expression is certainly simple enough, but we can simplify it even further by introducing another rule discovered by Newton, namely that a body of mass m, subjected to a force F, will accelerate at a rate given by

$$a = F/m$$

Now combining these two principles, if M is an unrestrained test mass then it will accelerate toward the body of mass m at a rate given by

$$a = \frac{Gm}{r^2} \quad (1)$$

If we measure m in grams and r in centimeters then

$$G = 6.672 \times 10^{-8}$$

and the acceleration is given in centimeters per second per second. It is the rate of change of velocity. For example, if we enter into the formula the radius of the earth (6.378×10^8 cm) and its mass (5.979×10^{27} grams) we find

$$g = 980.7 \text{ cm/sec/sec}$$

for the acceleration of gravity at the surface of the earth. It is the rate at which any material body picks up speed in falling when its constraints are removed. In recent years the unit centimeters per second per second has been given the name gal (short for galileo), so the nominal acceleration of gravity would be written as 980 gals.

Now let us move on to consider centrifugal force. This is the force exerted by a body upon its constraints when it is obliged to move in a circle. It is given by

$$F = mrw^2$$

where m is the mass of the body, w is its angular velocity*, and r is the radius of the circle. But here again it is more convenient to leave the mass of our test body unspecified and speak of

---

* Angular velocity is similar to the more familiar rotational velocity except that it is expressed in radians per second instead of revolutions per second—there being about 6.28 radians in one revolution of 360°.

The Tides

acceleration rather than force, That is,

$$a = F/m = rw^2 \qquad (2)$$

If we express r in centimeters then the result comes out in centimeters/sec/sec—namely in gals.

Now let us apply these lessons to the earth-moon system. Although the moon does indeed revolve around the earth one can observe more precisely that both bodies revolve about their common center of mass as shown in Figure 9. The distance between the two changes slightly during the month, but we can take as the nominal distance to the moon

$$d = 238\,850 \text{ miles} = 3.844 \times 10^{10} \text{ cm}$$

Now the distance to the common center of mass must satisfy

$$rM = m(d - r)$$

where $\qquad m = 0.0123\,M$

Therefore $\qquad r = 4.671 \times 10^8$ cm.

But we have for the radius of the earth

$$r_e = 6.378 \times 10^8 \text{ cm}$$

so r/r_e is less than one. In fact,

$$r/r_e = 0.732.$$

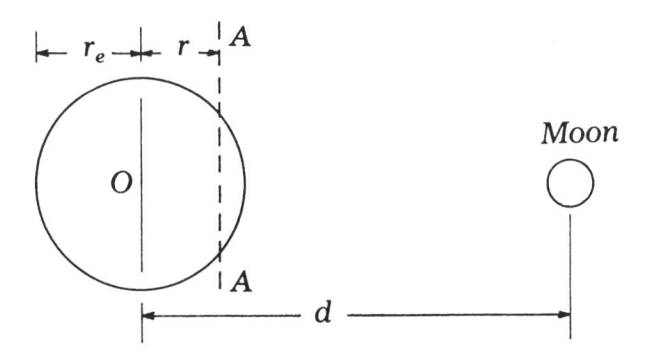

FIGURE 9: *A schematic representation of the earth-moon system where the line AA, passing through the common center of mass, represents the axis of revolution.*

Moreover, we know that the lunar period is about 27.322 days so the angular velocity of rotation is easily found to be

$$w = 2.6617 \times 10^{-6} \text{ radian/sec.}$$

Then using Equation (2) we can calculate the centrifugal acceleration of the earth's center that arises from its circular motion about the common axis AA. Directed away from the axis it is given by

$$a = rw^2 = 0.00331 \text{ gal.}$$

Likewise we can calculate the acceleration *toward* the axis which arises from the gravitational attraction of the moon. Putting $m = 7.352 \times 10^{25}$ grams for the mass of the moon we find from Equation (1)

$$a = \frac{Gm}{d^2} = 0.00332 \text{ gal.}$$

The two are very nearly the same, of course, the slight difference arising from the fact that the orbit is not quite circular. But let us keep in mind that these are the accelerations felt at the center of the earth. At points on the surface the cancellation will not be so exact, and the resulting imbalance is what gives rise to the tides.

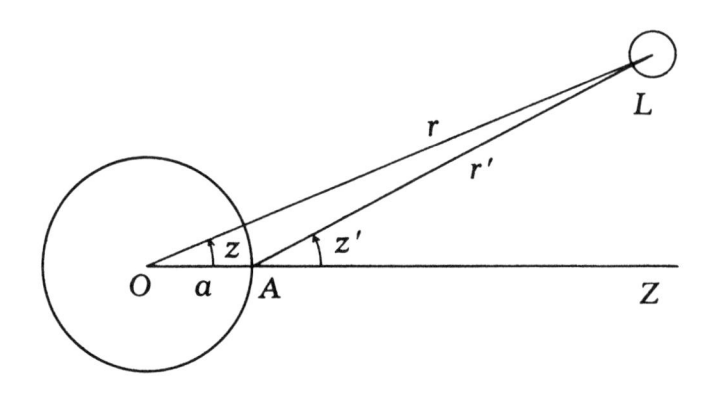

FIGURE 10: Schematic basis for calculating the tide-raising force. Here O represents the center of the earth, a is its radius, and A is a point of interest at the surface. OZ is the direction of the vertical at A, and L is an attracting body having mass m.

The Tides

In a recent work P. Melchior [59;p.13] reproduced the time-honored calculation for the tide-generating force. It proceeds as indicated in Figure 10. The problem is to calculate the difference between the acceleration of gravity felt at A and that prevailing at O, which, as we saw, is balanced out by the centrifugal effect. At O the component of acceleration along OZ will be

$$Z_O = \frac{Gm}{r^2} \cos z$$

whereas at A
$$Z_A = g - \frac{Gm}{r^2} \cos z'$$

and we need to calculate the difference between these two quantities. Melchior starts the reduction by noting three basic relations,

$$r' \sin z' = r \sin z$$
$$r' \cos z' = r \cos z - a$$

and
$$r'^2 = r^2 + a^2 - 2ar \cos z$$

the last being simply the law of cosines. The calculation is straightforward and leads to the result

$$t = Gm \frac{a}{d^3}(1 - 3 \cos^2 z) \qquad (3)$$

where terms of higher order in a/d have been neglected. Note that this expression gives a maximum (negative) value when the attracting body is either overhead ($z = 0$) or at the nadir, where $z = 180°$. Note also that at the angle z satisfying

$$\cos^2 z = 1/3$$

the difference vanishes, and between about 55 and 125 degrees away from the zenith the effect goes positive, thereby adding slightly to the strength of gravity. The maximum positive deviation occurs with the attracting body at the horizon where $z = 90°$, at which point it is just half as great as the maximum negative deviation.

Substituting our known values into Equation (3), and putting $z = 0$, gives for the maximum tide-driving acceleration arising from the moon

« 289 »

$$t_{-max} = -0.00011 \text{ gal}$$
$$= -110 \, \mu\text{gal}$$
where one μgal has been put equal to a millionth of a gal.

And we can just as easily calculate the departure from balance of the sun's attraction. For this case we would put
$$m = 1.991 \times 10^{33} \text{ grams}$$
and $\quad d = 1.4957 \times 10^{13}$ cm.
At $z = 0$ we find $\quad t_{-max} = -51 \, \mu$gals,
which is less than half of that contributed by the moon. Thus, the maximum departure from gravitational balance, prevailing when both bodies are near the zenith or nadir would be about
$$t_{-max \, total} = -161 \, \mu\text{gals} \qquad (4)$$
And correspondingly, with both bodies at the horizon we would expect a positive deviation of
$$t_{+max \, total} = +80 \, \mu\text{gals}$$

All of which appears to be in very good agreement with observation as we see in Plate 58. This is a portion of an actual gravimeter tracing taken at Brussels early in January when

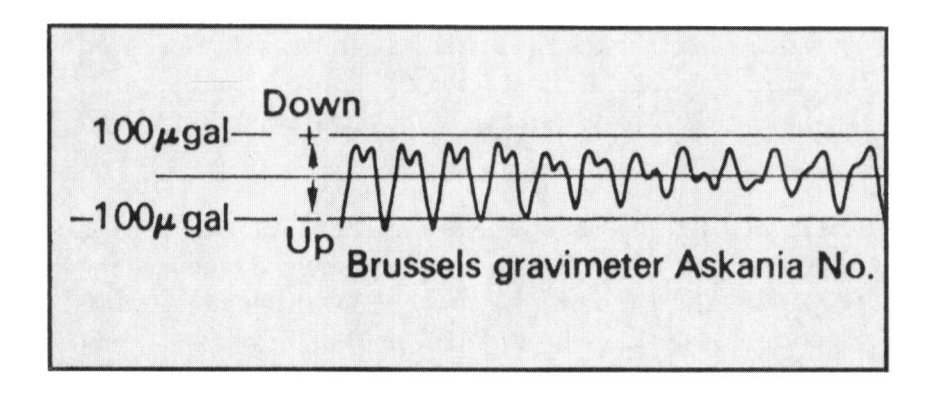

PLATE 58: *Segment of a gravimeter tracing made in early January at Brussels. The changing character of the plot is, of course, a reflection of the changing angle between the sun and the moon. (Reproduced from Page 18 of Reference 59)*

The Tides

both the sun and the moon crossed the meridian low in the southern sky. The observatory was so far north (latitude nearly 51°) that the deviation from normal gravity was positive during much of each day. We see that the positive deviation is indeed close to 80 μgals, and the negative dips, reaching about –110 μgals occasioned by not-so-near approaches to the nadir, also agree satisfactorily with the theory given above.

But what of the imbalance in centrifugal acceleration? Let us note first of all that since the sun is so far removed from the earth the r in Equation (2) is essentially constant over the entire earth. In that case the sun's gravitational attraction at the center of the earth very nearly balances the centrifugal force everywhere at the surface. But the moon presents a different situation entirely. Here the various points on the earth's surface vary greatly in distance from the axis of revolution. In fact, as we see in Figure 9, the point at the nadir is

$$1.732 / 0.732 = 2.36$$

times as far from the axis as the center of the earth so the unbalanced residue of centrifugal acceleration amounts to about
$$1.36 \times 0.00331 = 0.0045 \text{ gal}$$
$$= 4500 \, \mu\text{gals}$$

or nearly thirty times greater than given by Equation (4)! Here is a deeply disturbing riddle indeed because there is no proper reason for ignoring this unbalanced centrifugal acceleration in the theory of tides. It is quietly ignored, without ceremony or justification, simply because *Nature herself seems to ignore it.* But how can this be?

In the light of our new understanding a resolution to the riddle becomes readily apparent. Namely, we have seen that approximately half of the earth's gravitating strength stems from other worlds superimposed upon the earth and arrayed along that other dimension of space; presumably they are held in place simply by their mutual gravitational interactions. But then they would respond also to centrifugal acceleration so in fact their center of mass will not strictly overlie the center of the earth. Such a configuration would be unstable because one of lower potential energy is readily accessible. Namely, given such an arrangement initially, the two "half-earths" would proceed to separate. One half would "slide down the potential hill" toward the moon and the other half would slide down the other side. This separation would continue until the unbalanced centrifugal acceleration was balanced at last by the gravitational attraction between the two halves. With that condition realized the simplified theory given by Equation (3) becomes at least approximately valid*.

Let us note that if the centers of those worlds comprising earth's "other half" were constrained to overlap then the spacial distribution of gravitational acceleration would necessarily be different from the spacial distribution of centrifugal acceleration. In that case, even when the potential energy had attained its minimum value the two accelerations would not exactly cancel everywhere; a residual would usually remain. Presumably, however, the various centers are not constrained to overlap, but are free to move with respect to one another in such a way as to reduce their potential energy to its ultimate minimum. When this configuration had been achieved then the two forces would more nearly balance everywhere so departures from the theory of Equation (3) would be smaller yet. Nevertheless, it seems unlikely that they would ever vanish altogether so with sufficient care one should be able to observe systematic departures from that simplified theory.

Appendix B:

A Big Bang?

MOST OF THE BEST MINDS in the field of cosmology now agree that the universe is expanding and that the expansion began long ages ago as an initial "Big Bang". However, these same scientists also speak confidently of "Black Holes". Presumably these result when matter becomes so concentrated that even light cannot escape the resulting gravitational attraction. The mass required to form such a sink is relatively modest so the common understanding is that some of the larger stars will collapse as black holes when their nuclear fuel is exhausted. But the concept of black holes and an initial big bang appear almost mutually exclusive; one being valid would seem to preclude the other.

In fact, both concepts are open to serious question, having been born in an environment of only three dimensions. In this restricted realm a black hole is the logical consequence of gravitationally induced pressures so intense that they overcome the forces of exclusion that preserve atoms in their normal state. But given that added dimension of space it is not difficult to conceive of other means by which Nature might respond to such pressures. For one, the excess mass might simply "leak away", thereby allowing the remaining matter to retain its customary integrity. In this new light, therefore, black holes might well be illusory*. On the other hand the concept of an initial big bang is suspicious on its face, and might be abandoned gladly if the supporting evidence should admit of some other interpretation.

Let us pause to review this evidence. Superficially it is compelling indeed, being derived at least in part from spectroscopic studies of the light emitted by distant galaxies. When specific details in their spectra can be identified they are found to be shifted toward the red compared to those same details measured, say, in the spectrum of light from the sun. Classical physics recognizes only one mechanism for such a shift in wavelength; it is the Doppler effect which arises from relative radial motion between a source of light and its observer. A shift toward the red implies that the distance between the source and observer is increasing—and at a rate proportional to the fractional change in wavelength.

It turns out that the velocity at which galaxies recede from us, as deduced in this way, is at least approximately proportional to their distance. Obviously such distances cannot be determined to the same precision as the red-shifts. In practice, one has to assume that all galaxies have essentially the same intrinsic luminosity as our own milky way, and then the

---

Presumably the extreme pressure would cause the world "plane" to broaden until it overlaps the neighboring planes on each side, at which point the excess material would disappear from the one world and reappear in the others. It is interesting to note that this mechanism can also account for the loss of mass required in the formation of white dwarfs. These are thought to be the final stage in the evolution of larger stars—attained when their nuclear fuel has been exhausted. Remarkably enough the size of a white dwarf <u>decreases</u> with increasing mass, theoretically approaching zero radius for a mass about 1.2 times the mass of our sun [1;p. 587]. White dwarfs are fairly common and since they must have derived from stars substantially more massive than 1.2 solar masses the mechanism for the required loss of mass has long been a subject for speculation.

A Big Bang?

distance to a particular galaxy is deduced from its *apparent* brightness. Although this assumption cannot be strictly valid nevertheless it does yield a distinct correlation between this apparent distance and the amount of spectral shift toward the red. That is, the more remote galaxies suggest proportionately larger velocities, implying that their motions all started from a common point. Of course, this correlation is the essential foundation of the Big Bang picture.

Formerly those who studied this problem restricted their thinking to a space of three dimension, but we now understand that this constraint is unnecessary so let us enlarge our view and see what that extra dimension of space has to offer.

First of all one can deduce from its obvious properties that light itself must be a phenomenon is four dimensions. That is, on the one hand it manifests all the properties of a typical wave motion. Namely, it diffracts and shows interference effects that can be described in terms of a wavelength. But in other circumstances, as for example in the photoelectric effect, light displays particle-like behavior just as clearly. This dual character simply will not fit into any three-dimensional model, but it can be easily managed by making recourse to that extra dimension. We have only to recognize that a unit of light is neither a wavelet nor a particle, but is instead a "wavicle" in four dimensions. When viewing this entity we have to project it onto our three-dimensional space where it must appear either as a particle or a wavelet. It cannot display both properties at the same time because there is no such thing as a wavicle in three dimensions. Consequently, being an object in four dimensions, we can expect it to have a certain breadth along that added extension of space.

In comparing the universe to a tall apartment building in three dimensions we saw earlier that each of its "worlds", instead of being mere planes in two dimensions, had breadth along the third dimension. The analogy suggested that our own world in three dimensions might similarly have a certain breadth along the fourth. It seems safe to conclude that these two breadths—the breadth of light and the breadth of our world plane, exactly overlap. But what effect might one observe if the incident light did not overlap our world space precisely—if the light were displaced somewhat along that fourth dimension?

Obviously we have no basis for deducing *a priori* how this interaction might proceed so let us reason backwards and conclude that most of the galactic red-shift results from just such an offset between the incident light and our world-plane. Presumably only that portion of the incident quantum which actually overlaps our world-plane can be absorbed; the remainder continues on its way. Of course, this incomplete absorption of energy would manifest as a shift toward the red. Now let it be emphasized that this is a different mode of absorption than physics normally allows. That is, upon interacting with ponderable matter under normal circumstances a quantum of light is either entirely absorbed or entirely unaffected. Partial absorption of this kind is not observed in the laboratory because light generated there always overlaps the absorber exactly*.

One might suppose that some fundamental law of nature defines the separation between the various world planes, and evidently that separation is somewhat greater than the width of light because we are able to glimpse no more than one such plane at a time. Furthermore, we can deduce that the visible stars within our milky way galaxy all occupy the same world-plane since we observe only small red-shifts in their spectra, and those are easily identified as the Doppler shifts resulting from true proper motions.

But according to our picture the world-planes of widely separated, weakly interacting galaxies do not agree. Appar-

---

* *One might also note that this is a much different case than, say, shining light onto a partially smoked glass. Here individual quanta are either absorbed or transmitted on a statistical basis, but those that are transmitted retain their full energy.*

A Big Bang?

ently the more remote galaxies tend to be displaced progressively further from our own world-plane, causing their light to be shifted ever further toward the red. However, part of this correlation between distance and red-shift may be illusory because apparently an object displaced from our world-plane seems to shrink in size and therefore appears to be more remote than it actually is.

We gain this added bit of insight by observing the remarkable behavior of those enigmatical Unidentified Flying Objects. These craft exhibit any number of bizarre properties which suggest that they are somehow able to navigate in four dimensions. As a case in point, observers have watched these objects grow and shrink in apparent size even while hovering stationary in space.

Let us take a moment to view this spectacle through the eyes of an actual witness—a Captain Ferriera, pilot of an F-84 Thunderjet. He and the pilots of three other Thunderjets were on a training flight out of Ota Air Base in Portugal one clear night in 1957. The moon was almost full, and the visibility was more than fifty miles. The relevant portion of the Captain's account was relayed by Good as follows [39;p.147]:

> " ... The thing looked like a very bright star unusually big and scintillating, with a colored nucleus which changed color constantly, going from deep green to blue to passing through yellow and reddish colorations. ...
> " ... All of a sudden the thing grew very rapidly, assuming five or six times its initial volume, becoming quite a spectacle to see. ... [then] fast as it had grown, [it] decided to shrink, almost disappearing on the horizon, becoming a just visible, small, yellow point. These expansions and contractions happened several times, but without becoming periodic and always having a pause, longer or shorter, before modifying volume. The relative position between us and the thing was still the same, that is about 40° on our left, and we could not determine if the changing dimensions were due to very fast approaches and

« 297 »

retreats on the same vector or if the modifying took place stationary. ... After about seven or eight minutes of this the thing had been gradually getting down below the horizon and dislocated itself for a position about 90° to our left. ..."

Certainly this phenomenon has no place in our rational, three-dimensional world, but an obvious explanation comes to mind in light of that other dimension of space. Presumably it resulted from displacements of the craft back and forth along the fourth dimension while it remained otherwise at rest. Applying this lesson to our present problem we conclude that an offset along that extra dimension, in addition to causing a redshift, also renders the image of a galaxy smaller—thereby making it appear more remote than it actually is. Of course, this effect tends to undermine the presently recognized relationship between distance and red-shift. Indeed, in this light one might expect to observe comparably dim galaxies with widely different red-shifts.

It is interesting to note that this added dimming of remote galaxies offers an easy explanation for the fact that the sky is black. One might normally expect that the essentially unlimited number of stars in the universe would contribute a noticeable background glow to the night sky. For example, suppose that we divide all of space into a sequence of concentric shells, all of the same thickness, with ourselves at the center. Then each shell would contain on average a number of stars proportional to the square of its radius. But on the other hand the apparent brightness of those stars would *decrease* inversely as the square of that radius so on average each shell would contribute the same amount of light. But presumably the number of such shells is unlimited so some finite illumination should accrue however small might be the light provided by each individual shell. The failure of this reasoning can be traced to that progressive dimming of remote galaxies caused by the offset of their light from our own world-plane. Thereby

A Big Bang?

is the illumination arriving from successive shells not constant after all but decreases with distance—and evidently rapidly enough so that the accrued sum is very small.

This new insight also gives an easy resolution to the riddle posed by the so-called quasars, or quasi-stellar objects. These are relatively small light sources that also emit strongly at radio frequencies. Moreover, most of them are variable and display sudden changes in brightness—implying that they are very small compared to a normal galaxy. And most significantly of all they exhibit very large red-shifts, suggesting that they are extremely remote. In fact, they appear to be hardly more than stars, hence their name. But if one calculates their intrinsic luminosity from their apparent brightness and distance (as estimated from their red-shifts) they turn out to be prodigious objects indeed, as Abell explains [1;p.639]:

> " We have then the perplexing picture of a quasi-stellar source: an extremely luminous object of small size displaying enormous changes in energy output over intervals of months or less from regions less than a few light months across; 100 times the luminosity of an entire Galaxy is released from a volume more than 10^{17} times smaller than the Galaxy."

Such an object is so unlikely as to be incredible, but our new understanding of the red-shift offers an easy resolution to the puzzle. We recall from the footnote on page 294 that as a massive star collapses to become a white dwarf its mass must decrease to some value less than 1.2 times that of our own sun. There we supposed that the extreme pressure broadens the world-plane locally until it overlaps the neighboring world-planes, allowing any excess mass to leak away. Presumably this very hot material would reappear as a new star situated at the very edge of each of those two neighboring world-planes.

Now viewing this process from one of those neighboring planes, that new star (being situated at the very edge of the prevailing world-plane) would manifest an extremely large red-shift and would therefore be identified as a quasar. Its distance away, being deduced from that great red-shift, would be vastly overestimated so its intrisic brightness would be greatly overestimated as well. This model accounts for the very significant fact that the red-shift observed in every quasar is larger than in any known galaxy. One might go on to guess that the substantial radio emission from these objects results in some way from their gradual approach to the prevailing world-plane. Indeed, it is tempting to generalize and suggest that this is the mechanism behind all radio emission from celestial sources.

Of course, this would include the enigmatical pulsars also. These objects emit pulses of radiation in the radio-frequencey band with clock-like regularity, and some emit pulses of visible light as well. With these present findings in mind we might suppose that in some fashion these objects have been set into oscillation, moving back and forth along that fourth dimension. Certainly a physical oscillation of such massive bodies at these observed rates would be unthinkable, but no visible motion at all is implied here. Those that happen to be self-luminous would blink for the same reason that that U.F.O. blinked, as described by Captain Ferriera, and the pulsing radio emission would originate from the periodic *movement* along that fourth dimension.

Appendix C:

THE GEOMAGNETIC FIELD

WE HAVE SEEN that comets probably had a magnetic structure of some kind at their birth in the solar atmosphere so it is only a small step to suppose that the plasma issuing from them at their decay likewise supports an intrinsic magnetic field. Under normal circumstances in interplanetary space that field is completely dissipated when the plasma condenses, but in the presence of a neighboring conductive medium this circumstance might be altered. Namely, even after the plasma decayed back to the normal state of matter residues of its magnetic field might be preserved temporarily by electric currents *induced* in that nearby conductor. Here then, free for the taking, is a plausible accounting for the terrestrial magnetic field—which is doubly attractive in the absence of any other feasible model.

For as is well known the earth is not "magnetized" in the same sense as a permanant bar magnet. Its field derives from electric currents circulating in the hot, electrically conductive depths. These currents are presently thought to be generated by some kind of self-excited magnetohydrodynamic dynamo mechanism powered by thermal convection. However, in actual fact, many years of labored analysis have failed to show any such mechanism to be theoretically sound.

Of course, one could hardly hope to trace in detail the transition from a tightly contoured plasmoidal field to a greatly extended configuration that might induce large-scale circulating currents in the earth. However a start might be made simply by noting that the original plasmoid formed under the relatively high pressure prevailing in the solar atmosphere. Therefore one might expect that this same plasma, issuing into a near vacuum at the upper reaches of the terrestrial atmosphere, would tend to expand freely, expanding the field configuration as well. The field strength would therefore probably decrease substantially at the same time, but the *field energy* could easily increase at the expense of the thermal energy of the plasma. As the plasma thus cooled and condensed perhaps this larger-scaled but weaker field could be preserved temporarily by currents induced in the ionosphere. Then these currents decaying would induce long-lived currents in the earth itself. Probably the initial terrestrial field derived in this way would be a complex superposition of multipole components, of which the higher order components would die away most rapidly. That is, one would expect the dipolar component to predominate eventually because it can be supported by currents flowing entirely within the deepest—and hottest parts of the earth; presumably the medium is the most highly conductive there. Of course, this is the situation that prevails today; the dipolar component now makes up about 90% of the total field.

Let it be noted that this view of origins easily accommodates the fact that the dipole axis is tilted with respect to the axis of rotation and also that it misses the center of the earth by some hundreds of miles. The axis of rotation simply does not enter into the picture so there need be no connection whatever between the two directions. In fact, one can imagine that the dipolar component was even less symmetrically positioned in early times and that it has achieved its nearly central location through processes of decay and diffusion. Presumably we still

The Geomagnetic Field

see a hint of this diffusion today in the wandering of the magnetic poles. That is, as the circulating currents giving rise to the field diffuse down to ever greater depths they choose the paths of least resistance and thereby change their overall symmetry slightly. This (presently) slight change in current configuration would manifest as a corresponding shift in the position of the poles. The gradual westward trend of the non-dipolar component of the field is likewise readily explained. It results simply from the effective magnetic interaction between the various elementary dipoles making up the field; they seek the configuration of minimum field energy.

Observations during recent times have shown that the dipolar component of the field is presently decreasing at an appreciable rate. In fact, during the past century it has been dropping at an average rate of about 0.04% per year [16; p.164]. As seen from our present point of view this drop can be nothing but the normal decay resulting from the inherent resistance of the medium that carries those circulating currents. If this drop is interpreted as a segment of a simple exponential decay pattern it would imply a half-life for the dipole field of only about 1700 years. In that case our present terrestrial field could not be very old, and it should have been substantially more intense in the past than it is today—as much as six time as great even during historical times.

As it happens techniques have been developed for deducing the strength of the earth's magnetic field in antiquity from measurements of the remnant magnetization in bricks and other ceramics dating from those olden times. This slight magnetization stems from minute magnetic impurities in the clays that become partially aligned to the ambient field as a result of the firing process. These measurements tend to confirm the fact that the terrestrial field has been dropping for about the last 2500 years, although the rate of decay thus deduced is substantially less than would be expected from our present point of view. But more troublesome yet is the finding that at still earlier times the trend reversed, and the field seems to have *increased* with the passage of time. Such measurements suggest that earlier that about 3000 years ago the terrestrial field was much the same strength as it is today [60; p.105].

Taken at face value these results argue against our new point of view even though, strictly speaking, such determinations do not measure simply the dipolar component of the field; they include contributions from the various multipole components as well. At any one site this resultant field could well be weaker than that of the dipole field alone, but by the same token measured strengths at other localities should be abnormally high. However, if such high values have been observed the writer is not aware of them. Since this apparent inconsistency in the ancient field strength is the only obvious difficulty with our new view of the terrestrial field we must take the time to examine this measurement technique critically for possible sources of systematic error.

As already noted the remnant magnetization in those old clays stems from the presence of minute crystals of magnetite and related minerals. Such small crystals tend to be individually magnetized, and when these small magnets are preferentially aligned along a given direction, in a brick for example, then the brick itself will display weak magnetic properties. In the course of time some of these micro-magnets will align themselves to an ambient magnetic field—although not usually by physically turning. Instead the turning takes place at the atomic level, and having thus turned the new configuration will persist even after the external field is removed. Crystals which will change their magnetic orientation easily in this way, in response to weak external fields, are said to be magnetically "soft". Obviously such soft impurities in an ancient artifact confuse the issue when one attempts to determine the magnetization that was acquired anciently during the firing process. These soft crystals have long since "forgotten" the baking and reflect the field direction prevailing in the recent past instead.

The Geomagnetic Field

In order to circumvent this problem researchers have developed techniques for utilizing only the magnetically "hard" impurities in their determination of the ancient field. One typical technique works in this way: The magnetization of a sample is first measured in its naturally occuring state, and then the soft magnetization is removed in progressive stages. At each such stage the sample is placed in an alternating magnetic field of a prescribed initial maximum strength; then this alternating field is gradually reduced to zero, after which the surviving magnetization is measured anew. Of course, at each step the initial strength of this demagnetizing field is progressively increased so that eventually even the hardest crystals will be completely randomized in direction, and the sample will then show no net magnetization whatever. Ideally, when the surviving magnetization is plotted against the strength of that demagnetizing field at each stage the points gained near the end of the experiment will define a straight line. This line extended backwards to the beginning, then, defines the magnetization that would have obtained if all the grains in the sample had been equally very hard.

The next step in the analysis is to remagnetize the sample by heating it to firing temperature in an ambient field of known strength. When the sample has cooled in this field the foregoing sequence of measurements is exactly repeated to give a second sequence of data points. The same straight line extending backwards gives the magnetization that would have resulted from hard grains as before, but this time as generated by the known field. Then it is a simple matter to deduce the ancient field from the ratio between these two hard-grain magnetizations and the strength of that known field.

Now clearly, the fundamental assumption underlying this technique is that the total population of magnetically hard grains is the same in both cases. Or, what amounts to the same thing, one assumes that the various crystals are magnetically equally hard before and after that annealing procedure. However this assumption is open to question if we go one step further and ask why the direction of magnetization is normally stable in a small crystal. Namely, basic energy considerations select certain directions in the crystal as preferred for magnetization, and usually nature prefers one of these more than the others. This state of lowest energy, the one nature most prefers, is called the "easy" direction of magnetization. A well defined easy direction renders a crystal magnetically anisotropic and therefore hard. Now the degree of this anisotropy must depend on the degree of regularity of the crystal lattice. One can therefore anticipate that dislocations and imperfections that disturb the perfect regularity of the lattice would "homogenize" the crystal to some extent and render the crystal less anisotropic in its magnetic properties—that is, less hard, and in the extreme would undermine the magnetic properties altogether. Since we understand that irradiation by energetic particles can severely disrupt crystalline lattices it is tempting to suppose that the accumulated radiation damage inflicted by cosmic ray particles and local radioactive impurities over the centuries has been sufficient to undermine the validity of those deductions of the terrestrial field strength in antiquity. Because of this damage the small crystals are magnetically softer before the annealing step than afterwards.

It is interesting to note that techniques for measuring this accrued radiation damage have recently been developed as a tool for estimating the age of ancient ceramic artifacts. The method, called *thermoluminescence dating*, has been described in detail by Aitken [2]. It makes use of the fact that displaced atoms constitute, in effect, high energy states of a crystal, and when, through annealing, these atoms return to their proper sites in the lattice their excess energy may be emitted in the form of light. To perform the measurement the ceramic sample is first reduced to a powder which is then spread evenly over a thin tungsten plate that can be heated rapidly. Light emission begins at about 200°C and continues as the temperature in-

The Geomagnetic Field

creases until eventually it is swamped by the radiation of incandescence. Of course, the resulting luminescence is very weak and can only be detected with exceedingly sensitive instruments. Although many complications may arise which can undermine their accuracy as indicators of age these measurements do demonstrate the fact of physically significant accrued radiation damage in natural materials. One might hope that measurements of this kind could be used to correct for the magnetic softening effect noted above and thereby permit improved estimates of the geomagnetic field strength in antiquity, but let it be noted that this radiation damage amounts to the literal destruction of data so for very old samples the prospects for great improvements in accuracy by means of such corrections are dim.

As is well known, similar magnetization measurements are performed on lavas and prehistoric rocks in order to determine the *direction* of the magnetic field in olden times. But of course, this measurement requires only the first set of determinations described above; no remagnetization step is necessary so there is no obvious reason to question these findings. Such tests show that the terrestrial field has varied substantially in direction, even to the point of changing polarity completely. If the magnetohydrodynamic theory of origins is unable to account for even a steady field then it is vastly more troubled by this unlikely behavior. But such changes pose no problem whatever to our present point of view. Owing to its relatively fast decay the terrestrial field at widely different epochs must be traced to entirely different impact events, and of course there need be no connection whatever between the field configurations that might result from these different events. Moreover local field reversals on a much shorter time scale are also easily possible because, as we saw above, immediately following such an event the resultant field is a superposition of many multipole components which are not only irregular in space but also widely variable in time. That is, the higher multipole components would tend to decay first because the currents generating them cannot flow entirely within the hot depths of the earth where the resistivity is lowest. Therefore one can anticipate that the resultant magnetic field at any one site might change both in magnitude and direction several times following such an event before a stable dipolar residue would prevail. Unfortunately it follows that such measurements made on samples dating from that unstable period would be of only marginal significance, and the meaning of comparisons between widely separated samples would be even more doubtful.

Appendix D:

VOLCANISM

THE NEW INSIGHTS to the nature of comets derived in Chapter 8 of this study, and enlarged upon in Chapter 9, lead one to suspect that any improbable display of great energy might be in some way traceable to that same phenomenon. Because they evidence such enormous energies and manifest such high temperatures volcanoes must qualify for a reappraisal with this possibility in mind—especially as they often display features suggestive of the bizarre. In order to accomplish this review in a manageable space let us agree that the obvious properties of volcanoes are sufficiently well known that we need not rehearse them here, and neither will it be necessary to recount those peripheral details which do not relate directly to our problem. Confining our attention to merely a few specific properties, then, we shall be able to conclude that these eruptions are indeed long-lived residues of cometary encounter.

Let us begin by noting that volcanoes are relatively superficial disturbances. In the active state they are always accompanied by subterranean tremors and rumblings, the sources of which seismologists are able to trace with finesse. By far the majority of these tremors originate at depths of from one to four miles beneath the surface of the earth, and a plot of their origins tends to enclose a bulbous volume, from the interior of which essentially none originate. As Decker and Decker have explained at length the obvious implication of this distribution of origins is that the bulbous volume is in a fluid state and is the *immediate* source of the lava that is due to be ejected [26;p.113]. However the *ultimate* source is thought to be more deeply seated.

Namely, one needs to recall that temperature inside the earth increases steadily with depth. Measurements in deep mines and well shafts have shown that the rate of increase is typically about 48°C per mile of depth. Starting from a nominal surface temperature, then, one might expect to encounter temperatures comparable to the boiling point of water at a depth of a mile and a half or so, but the liquid rock in those bulbous chambers under volcanoes is more than a thousand degrees hotter than that. In fact, extrapolating that temperature gradient to greater depths one might not expect to find temperatures in excess of 1100°C for another twenty miles. According to current thinking, then, those chambers must be supplied by liquid rock forced up from below through long conduits.

In actual fact, this estimate of twenty miles may be unrealistically low—or in other words that relatively high temperature gradient may prevail only quite near the surface. The gradient presumably depends on the distribution of radioactive species in the native rock, the disintegrations of which are thought to be the source of this interior heat. It happens that these radioactive constituents are substantially more prevalent in the granitic crustal rocks than in the deeper lying basalts, suggesting that a smaller gradient must prevail beneath the crust. According to data supplied by the Deckers [p.164] best estimates are that temperatures inside the earth reach values near 1200°C only at depths approaching 60 miles! In that case we have to suppose that this melted magma is squeezed up through some sixty miles of narrow channels, through progres-

Volcanism

sively cooler rock without solidifying, to arrive at those shallow chambers—only then to manifest seismic effects. Of course in the absence of specific data about conditions in that deep realm one cannot rigorously disprove this picture, but it hardly seems plausible in any case.

Some have suggested that the new theory of plate tectonics (described on Page 256) offers a more plausible picture of volcanism. According to this model friction between the diving plate and the stationary one above would generate heat, melting the rocks locally. The developing pressure would presumably force this melt to the surface where it would manifest as a volcano. That such a suggestion should ever have been seriously offered testifies plainly to the difficulty of the problem posed by that other picture, but this alternative is inadequate as well. Most probably the heat generated in this hypothetical process would be conducted away before great temperatures could be realized, but in any case the heat would be generated along a *line of contact*, not at isolated points as we observe in volcanoes.

However, we find an easy alternative to both of these pictures in our new model for comets and the way they interact with planets. In Chapter 8 we found that comets are objects in four dimensions that originate at the sun. The material was thrown out from the sun into a special metastable state that provides no mechanism for the loss of thermal energy to outer space. Therefore, when the materials do return to our stable world-plane they do so with their original energy intact. Regardless of the actual size of a comet its materials return to our world-plane only in a narrow region, called the active eye. Furthermore, we found that material in that metastable state manifests very little mass so when such an object strikes a planet it survives intact. Its subsequent behavior and appearance then depend on the location of that eye in relation to the surface of the planet.

In particular, one would suppose that the active opening

on the comet that struck Venus was situated above the surface of the planet so that the evolving materials and their included thermal energy were retained in the atmosphere.

On the other hand, presumably those captured comets that discharge their materials close beneath the surface will eventually generate pressures sufficient to break through the surface and give rise to an eruption of the familiar kind. But one would anticipate a different result from those that discharge their load at still greater depths. Here the overlying burden of rock would inhibit such an eruption; the resulting strains would be more easily relieved by lateral movements of the surrounding rock or by deformations of the hotter, and therefore more pliable, mass below. Thereby does this picture lead one to expect exactly what is observed in earthquakes —namely, the sudden release of strain *at specific localized subterranean points* and not along fronts or the edges of plates as the prevailing theory would require. Moreover in this light one can understand why the most common emission from volcanoes is smoke and fine dust; these are the most obvious products of rapid condensation from a vapor phase. On the other hand it is not at all clear how magma from the depths could be continually ejected in this finely distributed form.

Now although the model seems simple enough one should not presume to lay down rules limiting the elemental processes that might take place beyond our own three-dimensional realm. That is, although the supposed cometary plasmas usually condense to form familiar rocky materials, nevertheless at other times sulphurous gases or even nearly pure water may result—as when superheated steam alone is observed to issue forcefully from volcanic cones. This great variability in the ejecta poses an obvious problem to the prevailing point of view, but there is ample room for such unlikely behavior within the scope of our new picture. Notice that this model does not identify the factors that govern the rate at which materials decay back to the prevailing world-plane. Evidently this rate is

Volcanism

subject to wide variation as well.

Another suggestion of the bizarre in volcanoes is found in the so-called *nouée ardente* (glowing hot cloud), first described during the eruption of Mount Pelée on Martinique in 1902. The spectre has since been reported by others who observed it from a distance, but what seems to have been its first modern occurrence was reported by no one. The cloud bore down upon the city of St. Pierre and reduced it to a flaming rubble in seconds; of the 30,000 inhabitants only two lone souls survived. Two months after that catastrophe had taken place trained observers were surveying the scene from a small boat offshore when apparently the phenomenon repeated, much the same as before. Those witnesses recounted their experience in these words [26;p.132]:

> " ... As the darkness deepened, a dull red reflection was seen in the trade-wind cloud which covered the mountain summit. This became brighter and brighter, and soon we saw red-hot stones projected from the crater, bowling down the mountain slopes, and giving off glowing sparks. Suddenly the whole cloud was brightly illuminated, and the sailors cried, 'The mountain bursts!' In an incredibly short span of time a red-hot avalanche swept down to the sea. We could not see the summit owing to the intervening veil of cloud, but the fissure and the lower parts of the mountain were clear, and the glowing cararact poured over them right down to the shores of the bay. It was dull red, with a billowy surface, reminding one of a snow avalanche. In it there were larger stones which stood out as streaks of bright red, tumbling down and emitting showers of sparks. In a few minutes it was over. A loud angry growl had burst from the mountain when this avalanche was launched from the crater. It is difficult to say how long an interval elapsed between the time when the great red glare shone on the summit and the incandescent avalanche reached the sea. Possibly it occupied a couple of minutes. ... Had any buildings stood in its path they would have been utterly wiped out, and no living creature could have survived that blast.

"   Hardly had its red light faded when a rounded black cloud began to shape itself against the star-lit sky, exactly where the avalanche had been. The pale moonlight shining on it showed us that it was globular, with a bulging surface, covered with rounded protuberant masses, which swelled and multiplied with terrible energy. It rushed forward over the waters, directly toward us, boiling and changing its form every instant. ... The cloud itself was black as night, dense and solid, and the flickering lightnings gave it an indescribably venomous appearance. ...

"   The cloud still travelled forward, but now was mostly steam, and rose from the surface of the sea, passing over our heads in a great tongue-shaped mass. ... Then stones, some as large as chestnut, began to fall on the boat. They were followed by small pellets, which rattled on the deck like a shower of peas. In a minute or two fine grey ash, moist and clinging together in small globules, poured down upon us. After that for some time there was a rain of dry grey ashes."

That spectacle goes well beyond mere violence and must lay to rest any lingering doubt about the essential nature of volcanism. The prevailing model cannot even begin to account for such odd behavior, but it is easily interpreted in the light of our enlarged point of view. Namely, that daunting apparition must have been a plasma in rapid, turbulent motion—one *almost* constrained to a limited volume by its own self-contained magnetic fields; namely, it was a plasmoid in the process of decay, a piece of the comet itself. As the ionized vapors condensed, forming a black mist of smoke-sized particles, sparking ensued—nature's way of preserving the electric currents generating those magnetic fields in the face of waning conductivity. Indeed, lightning is commonly observed in volcanic discharges, a fact which is only to be expected in the light of this picture. In its race down the mountain harboring tornado-strength winds the swirling plasma picked up rocky debris and became heavy with its burden. But when it reached the shore and began to lose its rocky load its density decreased, and thus

Volcanism

it began to rise in response to the buoying effect of the surrounding atmosphere. This, then, is how it happened that those witnesses in the boat happily survived to tell the tale.

Let us note that our model also resolves an obvious riddle that is seldom even mentioned openly, namely, that one volcano can erupt violently while others nearby remain quiet. *This indicates that the source of pressure giving rise to the explosion does not stem from the depths of the earth.* If the lava emitted by a volcano actually came from the fluid-like depths then that same pressure would manifest all over the world and would be relieved either by the crust of the earth expanding slightly or by an eruption from every active volcano on earth at the same time. The same principle must apply as to the braking system of an automobile. Pushing on the master cylinder increases pressure throughout the hydraulic system and actuates the brakes on all four wheels simultaneously. The solitary explosion of Mount St. Helens in 1980, as an example, is proof that the source of pressure was very near to the earth's surface and that hard, unyielding rock lay between that source of pressure and the depths of the earth.

Now as additional evidence for the essentially alien nature of volcanism one can point to various typical volcanic deposits for which no plausible source can be identified. As a case in point we might recall that Frank Hamilton Cushing compared (p. 78) the age of the Hohokam culture with a Zuni culture in New Mexico that had been overcome by a flow of molten lava. That lava bed lies mainly to the southeast of El Moro National Monument and covers some 500 square miles, yet it has no identifiable source whatever! The interaction with the Indian's house testifies that this was a relatively recent event so the idea that the normally expected volcanic peak had subsequently eroded away would be untenable. A similar kind of riddle is posed by an extensive deposit of typical volcanic ash in Antelope County Nebraska which similarly has no obvious source. This site lies just to the north and east of the sand hills shown in Figure 6, suggesting that the deposits had a common origin—along with the drift that begins only a few miles further to the east. Herds of wild animals were buried in the fall, and many of their skeletons have been excavated. A museum has been erected over one such dig where one can observe these residues actually in place. Evidently this otherwise ordinary volcanic material originated from above rather than from within the earth. Presumably such deposits must be traced to captured comets whose eyes were situated above ground level.

Since, as it appears, all comets originated at the sun their internal compositions ought to be substantially the same. And furthermore, if volcanoes are truly driven by cometary residues their ejecta should also be of essentially the same composition. But in fact volcanic materials vary widely in composition. Even materials ejected from the same volcano may change dramatically from time to time. If our reasoning has been correct thus far it must follow that when material enters that four-dimensional state *it no longer resolves into normal atoms and thereby loses its chemical identity**. For want of a better name let us dub the substance prevailing in this four-dimensional state "hypermatter". Subsequently, upon condensing back into ponderable matter, and having lost all memory of its former composition, its new compostion would presumably be determined by natural law—consistent with the variable conditions at hand. In that case even petroleum and natural gas might be identified as products of cometary decay; the potential energy residing in these chemicals could then be considered a residue of that thermal energy which originated from the sun.

---

* How thermal energy is preserved in that peculiar condition is surely not obvious, but let us not lose heart. There remains much about this phenomenon which is inherently beyond our understanding.

Appendix E:

IN RE. BIG BONES

SINCE THEIR FIRST DISCOVERY more than a century ago the mountains of bones scattered over the frozen wastes of Siberia have held the world in thrall. And they are not merely bones. Some whole animals were found that showed every evidence of quick freezing; they were so well preserved, it is said, that dogs would devour the flesh when it thawed. We are even told that the stomach of one of those old victims still contained the undigested remains of its last meal, and it consisted of flora found only in the tropics. Here is a riddle to test the faith of any rationalist for certainly no understanding for those old residues is possible that would be agreeable to common sense and reason. But rather than abandon their self-imposed constraints and seek understanding outside of the Laboratory, so to speak, scientists have preferred to let the mystery lie unresolved, professing to believe that a rational solution will be found eventually.

However, now that we understand the fact of an added dimension of space, by means of which nature may operate in ways that the reasoning mind cannot follow, we need not be so constrained any longer. By an easy extension of processes that the eyes of sober men have witnessed, we should be able to conclude with confidence that that great herd of animals fell victim to a Fortean-like event; namely, they precipitated into that frozen wasteland and undoubtedly died very quickly thereafter; presumably this fall, even as the fall of loess with its snails, was a consequence of the event of 1450 B.C. Thus do we avoid any thought of tropical conditions in those arctic regions or sudden changes in the weather that have to be contrived out of whole cloth. This picture also accounts for the huge number of victims—far more than the land could have supported even under ideal climatic conditions; one has only to cite as a precedent the fall of snails recounted in Chapter 7, where the tiny animals covered several acres like a layer of snow. And that was a relatively tranquil event generated only by a thunderstorm.

But of course many questions still remain unanswered —especially as concerns the source of the animals. Even granting that the snails, for example, fell from a cloud overhead their ultimate origin is a complete mystery. If they were "harvested" in some manner from a normal environment, whether in our world or another, then the bizarre reaping would have had to extend for miles in every direction. As an alternative one might suppose that the victims were somehow created in the process. That is, in such extreme circumstances perhaps nature has a way of generating multiple *real images* of existing objects—even of living animals. One might note in passing that a process of this kind would relieve the conceptual problem posed by the phenomenal amount of loess that fell on that occasion; we would no longer have to provide for a source of equivalent volume. Obviously this view violates everything we have understood about conservation laws, but then our mundane understanding is hardly relevant in this realm.

But in regard to the imagined cloning process, we cannot suppose the copies always to be exact duplicates because we saw that the lesser sand-eels described by Meek were not all of the same length. Implausible as such a process may seem at first sight, nevertheless, by a relatively modest extension, it

In Re. Big Bones

offers an easy resolution to still another riddle that has long confounded those who study old bones. This relates to the size and proportions of some of those fossil residues. The problem is not especially difficult to understand, but it does require some preliminary background so let us proceed to develop these points in an orderly fashion. To this end we shall find it useful to review J.E.I. Hokkanen's studies of the strength of bones and the "factor of safety" by which he measured the ability of an animal to sustain abnormal stresses [47].

Certainly one would normally anticipate that the mass of similarly shaped animals would increase in proportion to the cube of their linear size. And conversely, one would expect the linear Dimensions of corresponding bones to increase approximately as the cube root of an animal's Mass. That is, in a general way,

$$D = k M^{1/3} \qquad (1)$$

This would mean that the *cross-sectional area* of a particular bone would be proportional to the *square* of an animal's size or to the two-thirds power of its mass. Now directing our attention to the larger of the two bones connecting the knee with the ankle in the hind leg—the tibia, or shinbone, Hokkanen defined his safety factor as the ratio of the maximum load bearing capacity of this bone to the total weight of the animal. It would be therefore nominally

$$S = k M^{2/3} / M = k / M^{1/3}$$

Thus, as the animal's mass increases its safety factor goes down, approximately in inverse proportion to its linear size.

But more precisely, it turns out from actual measurements on many different animals that the tibia increases in diameter slightly more rapidly than the cube root of the weight as suggested above. Citing the work of Alexander and his associates Hokkanen gives for this dependence:

$$D = 4.8 M^d \qquad (2)$$

where D is the diameter of the tibia in millimeters, M is the animal's mass expressed in kilograms and d = 0.36 ± 0.02 instead of 0.33 as in equation (1). Using this more accurate relationship and the measured strength of bone itself Hokkanen's safety factor becomes:

$$S = 275 / M^{0.28} \qquad (3)$$

Over a range of values for M this factor turns out to be:

M	S
50	92
500	48
5000	25
50000	13
140000	10

Of course the mass of an extinct animal, as it had been in life, is not so easy to determine, but two different techniques are employed to estimate it. The one favored by Alexander [3] makes use of a scale model conformable to the reconstructed skeleton. The volume of this model is determined by measuring the amount of water that it displaces, and then the mass of the animal itself is derived by multiplying this volume by the cube of the scale factor and guessing at the original density. The greatest ambiguity arises from the necessarily somewhat subjective construction of the model. But note that even modern animals of the same height and skeletal form may differ considerably in weight. In this light, then, those different estimates of weight could all be truly representative of actual animals.

The technique for estimating weight favored by Anderson et al [4] specifies certain measurements to be made on the bones of those extinct animals, and then their original weight is deduced by comparing those measurements with similar data taken from present-day mammals. As a basis for such deductions they measured the circumference of the femur and humerus of different animals of various weights in order to derive a general functional relationship. They find that

In Re. Big Bones

$$M = 0.000084 \, C^{2.73} \qquad (4)$$

where M is the body mass in kilograms and C is the combined circumference of the thigh bone and the upper arm bone*. Of course the question that one might now ask is how reliably this relationship can be extrapolated from the regime of small animals into the realm of giants. A hint of a problem here is already to be found in applying this rule to the case of a large elephant. As Alexander points out [3;p.24] the measured circumferences of those bones from a 5900 kilogram elephant imply a weight by equation (4) of 9000 kilograms. That is, its bones are *larger than expected* in the ratio of

$$(9000 / 5900)^{1/2.73} = 1.167$$

Now let us recall Hokkanen's safety factor and the list of values derived from equation (3). We have just seen that the bones of the elephant are actually larger than would be deduced by extrapolating from measurements on smaller animals. From equation (3) one would expect a 5900 kg elephant to have a safety factor of slightly over 24—its tibia should support 24.2 times its body weight without crushing. But this animal's bones are larger than expected and should sustain even more weight than that. In fact, in view of the datum above its safety factor would be larger by a factor of

$$(1.167)^2 = 1.36$$

Namely, its safety factor would be closer to 33. Could this "bonus of bone" given to the elephant be Nature's statement of a lower limit to the safety factor for viable animals in the wild? Hokkanen did not think so; he assumed a lower limit of 10 for this factor instead, presumably in order to accommodate the

---

* *Note that the reciprocal of this exponent, 1/2.73 (=.366) agrees with Hokkanen's exponent in equation (2). This is not entirely trivial since Hokkanen was considering the tibia while Anderson et al measured the combined circumference of the humerus and femur.*

heaviest of the known dinosaurs.

Largest among the dinosaurs were various species of Sauropodomorphs, familiar for their bulky torso, long tail, and a remarkably long neck topped by a smallish looking head. Judging from their surviving fossil skeletons some of them were truly gigantic. One of these, *Brachiosaurus brancai*, is the largest complete skeleton ever assembled. Its head rises some thirty-nine feet above the floor, and it would have weighed in life, by various estimates, between 47000 and 87000 kilograms, nominally ten to twenty times the weight of a large elephant.

Now we saw that the bones of an elephant are substantially larger than one would expect from equation (2), which results in a safety factor more than a third larger than given by equation (3). Contrarily, as it happens, the bones of *Brachiosaurus brancia* are actually *smaller* than given by extrapolating values from the smaller animals. Or, put another way, the weight of that giant as estimated from equation (4), using the measured circumferences of the femur and humerus, turns out to be *substantially less* than estimates derived from its volume. As a case in point, using the volumetric method Colbert deduced a mass of 87000 kg for *Brachiosaurus*; by similar means Alexander found 46600 kg, but Anderson calculated only 31600 kilograms using equation (4), fully a third less than even the most conservative result by the volumetric method [3;p.25].

Now let us translate this finding into an effect on safety factor using equation (3) which also extrapolates from measurements on smaller animals. Namely, inserting 31600 kilograms into (3) we would expect a safety factor of 15. But using a minimum weight of 46600 kilograms which the size of the skeleton as a whole would suggest results in a safety factor of only 10, *a mere third of that which Nature thought best to bestow upon the elephant!* And in some other species the problem is even more acute. For example, by the volumetric method Alexander estimates the mass of *Diplodocus carnegiei* as 18500 kilograms whereas Anderson *et al* calculate from equation (4)

In Re. Big Bones

only 5800 kilograms, comparable to the mass of a elephant. But the skeleton of that huge beast as a whole shows the animal to have been much larger than an elephant; the problem is that *its individual bones are simply too small for its size.* If the 18500 kilogram estimate is correct then the animal had a safety factor built into its bones of only 7.6! In short, if elephants need that extra thickness of bone in order to be viable then those huge dinosaurs were wholly unsuited for life in the wilds. They would have been excessively vulnerable to broken bones from even the slightest mis-step.

If more evidence of a fundamental conflict were needed it can be found in the problem of supplying blood to the brain of such great animals. To put the difficulty in perspective we might consider the tallest of the living animals—the giraffe, which can attain a height of about 18 feet. With this animal a fair share of its height derives from its long legs so the position of its heart is about midway between the head and ground. That puts the brain about 9 feet above the heart so an exceptionally high blood pressure is required in order to supply the head with this vital fluid. It is remarkable indeed that its arteries are able to sustain such a high pressure without rupturing. But *Brachiosaurus* stood more than twice as high as the tallest giraffe, and moreover its heart was scarcely more than a third of the way between the ground and head when the head was raised to its maximum. In that condition the brain was close to 27 feet above the heart so that gigantic pump would have had to supply blood at a pressure *three times* that even in a giraffe! Its entire vascular system would have been vulnerable.

Without a doubt we have a serious problem with those huge animals that needs fixing, and one possible resolution may be found in those other worlds that we discovered overlapping our own along that other dimension. We found that they were of varying sizes, and if one of them had, say, a third of the earth's diameter then its gravitational field at the surface would also be down to about a third of the value prevailing here on earth. If such a planet had been the original home of *Brachiosaurus brancai* then the safety factor built in to that very same tibia would have been three times as large, about 30, not much below that of an elephant here on earth. At the same time the blood pressure required of its heart would drop by a factor of three, being then just equal to that for a giraffe.

Although this possibility would relieve the static stress on the bones of *Brachiosaurus* it leaves the dynamic stresses largely unchanged. That is, a 47 ton animal in motion represents the same momentum on our supposed mini-planet as it does here on earth so the hazzards to the bones inherent in starting, stopping and slueing would be much the same. Moreover the diet required to sustain such a huge animal would be enormous, and it is problematical whether adequate nourishment could be crammed through its relatively small head even if it were mobile and could eat continuously around the clock.

The remarkable winged reptile *Pteranodon* illustrates this problem in another context. It seems to have been a bat-like animal weighing about forty pounds and having delicate membranous (not feathered) wings spanning some 23 feet! Desmond [28] has described at length the many problems that that ungainly creature would have faced in life. For one, although it would have been supurb as a soaring glider it lacked the musculature to be able to take off under its own power, and in fact would have been only marginally capable of powered flight at best. As he explains (p. 182):

> " It would be a grave understatement to say that, as a flying creature, *Pteranodon* was large. Indeed, there were sound reasons for believing that it was the largest animal that ever *could* become airborne. With each increase in size, and therefore also weight, a flying animal needs a concomitant increase in power (to beat the wings in a flapper and to hold and manoeuvre them in a glider), but power is supplied by muscles which themselves add still more weight to the structure. ... There comes a point when the weight is just too great to permit

In Re. Big Bones

the machine to remain airborne. Calculations bearing on size and power suggested that the largest weight that a flying vertebrate can attain is about 50 lbs: *Pteranodon* and its larger but lesser known Jordanian ally *Titanopteryx* were therefore thought to be the largest flying animals"

That was before the skeletons of three ultralarge pterosaurs were unearthed in Texas which dwarfed even *Pteranodon*. They are incomplete but the upper arm bone of this giant was fully twice the length of *Pteranodon's*, giving it an estimated wing span of some fifty feet—and it would have weighed close to 400 pounds!

An obvious way around this embarrassing impasse is to recognize, as in Chapter 12, that those creatures appearing as fossils were victims of a Fortean-like event precipitated by the impact of that great comet. From the great numbers of falling victims in "normal" Fortean falls we are led to admit the possibility that *Nature has some way of generating multiple real images of a given original object—even of living animals*. But if she does indeed have this bizarre capacity then it is only a small step to suppose that those real images might be *enlarged from the original*. How this might be accomplished is certainly not possible to comprehend, but in the realm of the frankly bizarre one can scarcely hope to understand intricate details. The best that he can desire is to recognize gaps in the domain of reason when he comes upon them.

Supposing, then, that *Brachiosaurus* had been thus magnified by a factor of three he would have stood only 13 feet tall on his home planet, and his blood pressure would have been comparable to that of a giraffe even if gravity on that planet were equal to its force here on the earth. Furthermore, its mass at home would have been only about 1700 kilograms so the safety factor built into its tibia would have been about 30 as before. Thus if in truth Nature does have a means for generating multiple real images of an object, and if those images can indeed be enlarged, then this mechanism provides an easy solution to many problems associated with the great size of some dinosaurs, and it relieves the strain on the food supply at the same time; the animals would have been of tractable size with manageable appetites in their home environment, and *Pteranodon* might indeed have been originally comparable to a bat. Also in this light perhaps one can understand why the bones of *Diplodocus carnegiei* are so spindly; he may have been so small an animal at home that he did not qualify for a bonus of bone like the elephant and then he was *magnified* to a large size following equation (1) instead of *growing* to a large size in keeping with equation (2).

Remarkably enough Holden [48] presents compelling support for this conclusion while defending a different model entirely. Namely, he points out that the strength of a muscle group must generally increase in proportion to its cross-sectional area—that is, to the square of an animal's size. On the other hand, an animal's mass must increase generally as the cube of its size. Thus, the loading on an animal's musculature must increase in proportion to its size. Consequently Nature herself defines a limiting weight for any animal—that is, *when its muscles are just capable of supporting its own weight,* and Holden finds this limiting weight to be about 21,000 pounds, *just slightly more than the weight of a large elephant.* Thus, in keeping with Holden's conclusion, *those huge animals could not have survived here on earth.*

For completeness one should acknowledge that an altered force of gravity on those other planets could have played a part in proportioning the bones that we find today as fossils, but of course we have no way of taking that effect into account here.

REFERENCES

1: ABELL, GEORGE, *Exploration of the Universe*, Second Edition Holt, Rinehart and Winston, 1969.
2: AITKEN, M.J., *Thermoluminescence Dating*, Academic Press, London, 1985.
3: ALEXANDER, R. MCNEIL, *The Dynamics of Dinosaurs and other Extinct Giants*, Columbia University Press, 1989.
4: ANDERSON, J.F., A. HALL-MARTIN, and D.A. RUSSELL, Long bone circumference and weight in mammals, birds and dinosaurs, *Journal of Zoology (A)* **207**: 53 - 61 (1985).
5: BAILEY, M.E, S.V.M. CLUBE, and W.M. NAPIER, *The Origin of Comets*, Pergamon Press, 1990.
6: BALDWIN, PERCY M. (translator), *Discovery of the Cities of Cibola*, Fray Marcos de Niza. Hist. Soc. of New Mexico Publ. in Hist.,V.1., Albuquerque; El Palacio Press,1926.
7: BARTLETT, KATHARINE and HAROLD S. COLTON, A note on the Marcos de Niza inscription near Phoenix, Arizona, *Plateau*, **12**: 53 - 59 (1940).
8: BEAUMONT, COMYNS, *The Mysterious Comet*, Rider & Co., London, 1932.
9: BEAUMONT, COMYNS, *The Riddle of Prehistoric Britain*, Rider & Co., London, 1945.
10: BENHAM, JAMES W., Map of Salt River Valley, Arizona showing

the location of Ancient Canals and Cities, Phoenix Free Museum, 1903.
11: BERG, L.S., The origin of loess,
Gerl. Beitr. Geophysik **35**: 130 - 150 (1932).
12: BILLINGS, MARLAND P. , *Structural Geology*, Prentice-Hall, Englewood Cliffs, New Jersey, 1954.
13: BLACKER, I.R. and HARRY M. ROSEN, *The Golden Conquistadores*, Bobs-Merrill Co., Indianapolis, 1960
14: BOLSIGER, H., H. LECHTIG, and J. GEISS, A close look at Halley's comet, *Scientific American* **259**: No.3 (September 1988), pp. 96 - 103.
15: BOLTON, HERBERT EUGENE, *Kino's Historical Memoir of Pimería Alta*, The Arthur H. Clark Co., Cleveland; 1919.
16: BOTT, MARTIN H.P., *The Interior of the Earth*, Edward Arnold, London, 1971,
17: BULLARD, FRED M., *Volcanoes of the World*, University of Texas Press, Austin, 1984.
18: BURKE, REV. JAMES T., *This Miserable Kingdom*, Christo Rey Church, Santa Fe; 1973.
19: BURRUS, EARNEST J. (translator), *Kino Writes to the Duchess, Letters of Eusebio Francisco Kino, S.J. to the Duchess of Aveiro*, St. Louis University, St. Louis; 1965.
20: BURSTEIN, DAVID and VERA C. RUBIN, The distribution of mass in spiral galaxies, *Astrophys. J.* **297**: pp. 425 - 435.
21: CARDONA, DWARDU, Intimations of an Alien Sky, Aeon, Vol. I, No. 4 (1988).
22: CARDONA, DWARDU, Darkness and the Deep, Aeon, Vol. III, No. 3 (1990).
23: CARMACK, ROBERT M., *The Quiché Mayas of Utatlán*, The University of Oklahoma Press, Norman, 1981.
24: COURVILLE, DONOVAN A., *The Exodus Problem and its Ramifications*, Challenge Books, Loma Linda, Calif., 1971.
25: CRUIKSHANK, D.P., The development of studies of Venus, in *Venus*, D. M. Hunten, L. Colin, T.M. Donahue and V.I. Moroz, Eds., Univ. of Arizona Press, Tucson, 1983.

References

26: DECKER, ROBERT and BARBARA DECKER, *Volcanoes*,
W. H. Freeman Company, New York, 1989.
27: DELORIA, VINE, JR., *Red Earth, White Lies*,
Scribner, New York, 1995.
28: DESMOND, ADRIAN J., *The Hot Blooded Dynosaurs*,
The Dial Press, New York, 1976.
29: DONNELLY, IGNATIUS, *Atlantis and the Antedeluvian World*,
Harper and Brothers, New York, 1882.
30: DONNELLY, IGNATIUS, *Ragnarök: The Age of Fire and Gravel*,
(reprint of 1883 edition), Rudolf Steiner Publications,
Blauvelt, New York, 1971.
31: ENGLERT, FATHER SEBASTIAN, *Island at the Center of the World*, Charles Scribner's Sons, New York; 1970.
32: FLAMMARIAN, CAMILLE, *The Flammarian Book of Astronomy*,
Simon and Schuster, New York, 1964.
33: FLINT, RICHARD FOSTER, *Glacial and Quaternary Geology*,
John Wiley and Sons, New York; 1971.
34: FORT, CHARLES, *The Book of the Damned*,
Ace Books New York (paperback).
35: GEIKIE, JAMES, *The Great Ice Age*,
D. Appleton and Company, New York, 1888.
36: GENTRY, ROBERT V., *Creation's Tiny Mystery*, 3rd Edition,
Earth Science Associates, Knoxville, 1992.
37: GILLULY, JAMES, AARON C. WATERS and A.O. WOODFORD,
Principles of Geology, 3rd Edition, W.H. Freeman and Company, San Francisco 1968.
38: Gloucester HERALD (date not given) quoted in *The Philosophical Magazine* **58**: 310 - 311 (1821).
39: GOOD, TIMOTHY, *The World-wide UFO Cover-up*,
Quill, William Morrow, New York, 1988.
40: HALLENBECK, CLEVE, *The Journey of Fray Marcos de Niza*,
Greenwood Press. Westport, Connecticut; 1949.
41: HANCOCK, GRAHAM, *Fingerprints of the Gods*,
Crown Publishers, Inc. New York, 1995
42: HAPGOOD, CHARLES, H., *Maps of the Ancient Sea Kings*,

Chilton Company, Philadelphia, 1966.
43: HAURY, EMIL W., *The Hohokam, Desert Farmers and Craftsmen; Excavations at Snaketown 1964 - 1965.* University of Arizona Press. Tucson; 1976.
44: HAWKINS, GERALD S., *Stonehenge Decoded,* Doubleday & Company, Inc. New York, 1965.
45: HAWLEY, F.G., The manufacture of copper bells found in southwestern sites, *Southwestern J. of Anthropology* Vol **9**: pp 99 - 111 (1953).
46: HIGGINS, GODFREY, *Anacalypsis,* (reprint of 1833 edition) University Books, Hyde Park, N.Y. 1965.
47: HOKKANEN, J.E.I., The size of the Largest Land Animal *Journal of Theoretical Biology* **118**: 491 - 499 (1986).
48: HOLDEN, THEODORE A., Dinosaurs and the Gravity Problem, *The Anomalist* **1**: No. 1, 6 - 19 (1994).
49: HOWORTH, H.H., The loess—a rejoinder, *Geol. Mag.* **9**: 343 - 356 (1882).
50: KAKU, MICHIO, *Hyperspace,* Oxford University Press, New York, 1994.
51: KARNS, HARRY J.(translator), *Unknown Arizona and Sonora,* from *Luz de Tierra Incongnita* by Captain Juan Mateo Manje, Arizona Silhouettes, Tucson, 1954.
52: KEILHACK, K., Das Rätsel der Lössbildung, *Deut. geolog. Gesell. Zeit.* **72**: 146 - 161 (1920).
53: KINO, EUSEBIO FRANCISCO, S.J., *Exposicion Astronomica de el Cometa Que el Año de 1680, etc.,* Mexico City, 1681.
54: KURTZ, PAUL, Ed., *Humanist Manifestos I and II,* Prometheus Books, New York, 1973.
55: LUGN, A.L., The origin of loesses and their relation to the Great Plains in North America, in C.B. Schultz and J.C. Frye, Eds., *Loess and Related Eolian Deposits of the World,* University of Nebrasky Press, Lincoln, 1968.
56: MARSDEN, BRIAN G., *Catalogue of Cometary Orbits, 4th Ed.* Smithsonian Astrophys. Observatory, Cambridge, 1982.
57: MAZIERE, FRANCIS, *Mysteries of Easter Island,*

References

W.W. Norton & Co., Inc., New York, 1968.

58: MEEK, A., A shower of sand-eels, *Nature* **102:** (1918).

59: MELCHIOR, PAUL, *The Earth Tides,* Pergamon Press, Oxford, London, Etc., 1966.

60: MERRILL, R. T. and M.W. MCELHINNY, *The Earth's Magnetic Field,* Academic Press, Inc. London, 1983.

61: MÉTRAUX, ALFRED, *Easter Island,* Oxford Univ. Press, 1957.

62: MONTY, SHIRLEE, *May's Boy,* Thomas Nelson Publishers, Nashville, 1981. See also JOSEPH BLANK, The Miracle of May Lemke's Love, *Reader's Digest,* October 1982, p. 41.

63: MORRISON, DAVID, Voyages to Saturn, NASA Headquarters, NASA SP-451 (1982). (As cited by Richard Hoagland in a posting on the Internet.)

64: OORT, J.H., Empirical data on the origin of comets, in Barbara M. Middlehurst and Gerard P. Kuiper, Eds. *The Moon Meteorites and Comets,* Univ. of Chicago Press, 1963.

65: PATRICK, H.R., The Ancient Canal Systems and Pueblos of the Salt River Valley. The Phoenix Free Museum, 1903.

66: POSNANSKY, ARTHUR, *Tiahuanacu: Cradle of American Man,* (4 Volumes), J.J. Augustin, New York, 1945.
As cited by Graham Hancock.

67: PRIEST, JOSIAH, *American Antiquities and Discoveries in the West,* Hoffman and White, Albany, 1834.

68: RICHTHOFEN, BARON F., On the mode of origin of the loess, *Geol. Mag.* **9:** (1882) 293 - 305.

69: RUHE, ROBERT V., *Quaternary Landscapes in Iowa,* Iowa State University Press, Ames, 1969.

70: San Francisco EXAMINER, Jan. 22, 1888. See also issues of Nov. 20 and Dec. 25 of 1887 and Jan. 1 of 1888.

71: SAUER, CARL, *The Road to Cibola,*
The University of California Press, Berkeley, 1932.

72: SHEAHAN, J.W. and G.P. UPTON, *The Great Conflagration* Union Publishing Co., Philadelphia, 1871.

73: SIMPSON, GEN. J.H., Coronado's March in Search of the Seven Cities of Cibola & Discussion of their probable

location, *Smithsonian Institution Annual Report for 1869*.
74: SKERTCHLY, S. B. J. and T. W. KINGSMILL, On the loess and other superficial deposits of Shantung (North China), *Quart. J. Geol. Soc. (London)* **51**: 238 - 254 (1895).
75: SKJÖLSVOLD, ARNE, The stone statues and quarries at Rano Raraku, in Heyerdahl, Thor and Edwin Ferdon, Eds., *Archaeology of Easter Island*, Forum Publishing House, Stockholm, 1961.
76: SMITH, CARLYLE, The Poike Ditch, in Heyerdahl, Thor and Edwin Ferdon, Eds., *Archaeology of Easter Island*, Forum Publishing House, Stockholm, 1961.
77: STEEDE, NEIL, *Cro-Magnon Civilization*, To be published.
78: THOMAS, J. and P. VOGEL, Testing the Inverse-Square Law of Gravity at the Nevada Test Site, *Phys. Rev. Lett.* **65**: pp. 1173 - 1176 (1990).
79: TURNEY, OMAR A., Prehistoric Irrigation, II, *Ariz. Historical Review* **2**: 11 - 52 (1929).
80-: VELIKOVSKY, IMMANUEL, *Ages in Chaos*, Doubleday & Company, Inc., Garden City, New York 1952.
81: VELIKOVSKY, IMMANUEL, *Worlds in Collision*, Doubleday & Company, Inc., Garden City, New York, 1950.
82: WILLIS, B., E. BLACKWELDER and R. H SARGENT, *Research in China*, Vol. 1, Carnegie Inst., Washington, D.C., 1907.
83: WINCHELL, N. H., The geology of Rock and Pipestone Counties, *Ann. Report Geol. and Nat. Hist. Survey of Minnesota* **6**: 93 - 111 (1878).
84: WURM, K., The Physics of Comets, in Barbara M. Middlehurst and Gerard P. Kuiper, Eds. *The Moon, Meteorites and Comets*, University of Chicago Press. Chicago 1963.

INDEX

Active eye, *see comets*
Acus, *see Cibola, Cities of*
Ahacus, *see Cibola, Cities of*
Ahu Akivi 222, 223
Amber 175, 185
Angular momentum 178, 179, 180, 240, 243 - 246
Angular momentum space 179
Angular velocity 286
Anomalies, magnetic 257, 258
Antarctica 232, 245
Athena 189
Atlantis 234 - 238, 246, 255
Autistic savant syndrome 267
Barnes Butte 105, 111, 112, 201, 203
Beaumont, Comyns 124, 175, 181, 236, 255 - 257
Benham. James, W 60, 80, 87 - 91, 96, 99
Berg, L. S. 140, 143, 145, 146, 267
Big Bang 293 - 300
Black Holes 293
Black sky 298
Boat houses 215 - 217
Bone, strength of 319 - 321
   safety factor in 319 - 322

Boulder clay 120, 121, 122, 143, 168
Cabeza de Vaca *see Nuñez, Álvar*
Camelback Mountain 98
   melted rocks on 114, 115, 197
Casa Grande 52 - 58, 73, 74, 196
Cascabells 30
Castañeda, Pedro de 37, 38
Castillo Maldonado, Alonzo del 9
Centrifugal force 286
Chalk deposits 255
Chicago fire 193 - 195, 227
Chichilticale 35, 36, 37, 42
Cibola, Cities of 7, 8, 16 - 29, 31- 43, 50, 62, 70 - 74, 79 82, 85, 196, 205, 207 - 210 227, 244, 277, 280
   Acus 17, 22, 28, 209
   Ahacus 21, 22, 79
   Marata 17, 21, 22, 28, 74, 83, 209
   Totonteac 17, 19, 20, 22, 27 28, 58, 74, 79, 193, 209
Comets 120, 174 - 180, 191, 195 208, 229, 236
   active eye 181, 185, 191, 311
   angular momentum of

177 - 180
axial structure of 181
composition of 270
dirty snowball model
176, 177, 187
Donnelly's picture 176
Effective mass of 182
new model 185
origin of 178, 180 - 181
Comet Biela 194, 307
Comet Halley 177, 182, 244
Comet of 1680 207 - 209
Compostella 12, 29, 31, 41
Continental drift 257, 258, 311
Coronado, Francisco Vázquez de
8, 12, 13, 16-29, 31-43,
50, 62, 72, 79, 85, 196,
207, 210, 227, 244
Council Bluffs, Iowa 127, 129,
144, 145, 152, 153
Courville, D. A. 248
Culiacán 12, 13, 29, 32, 33, 35
40, 41
Cushing, Frank Hamilton 59,
76, 77, 80, 85, 86, 315
Diaz, Melchior 33, 35, 51
Dimension, added of space 163
173, 282 - 284, 291, 293 - 300
Dinosaurs 254, 256, 322 - 325
Diodorus Siculus 231
Donnelly, Ignatius 120, 122,
123, 124, 174- 176, 181
192, 193 - 195, 235, 257
Dorantes, Andrés 9, 10
Drift, glacial 120, 122, 124, 125,
151, 171, 174, 175, 257
Earthquakes 258, 311, 312
Easter Island 212 - 228, 279 - 282
giant statues on 212 - 215,
226, 277 - 280
top-knots on statues 215
Ecliptic 240
obliquity of 240, 242, 243
247, 248

Egyptian chronology 191, 194
239, 247, 248
intermediate periods 248
El Morro National Monument
66, 68, 313
Englert, Sebastian, Father 216,
220
Enlightenment 2, 249, 263, 264
269 - 276
Enlightened view of man 264, 271
Equinoxes 240
precession of 282, 283
Estéban de Dorantes 9, 10,
13-30, 34, 37, 39, 50
Evolution, theory of 256, 263,
275
Exodus out of Egypt 192, 237, 248
Flint, Richard Foster, 120, 121,
122, 130, 152, 266
Fort, Charles 165
Fortean events 168, 169, 171, 183
186, 250, 262, 281, 318, 325
Fossils, formation of 249 - 253
jellyfish 251, 252
deposition of 253 - 257
Galaxies, invisible mass in 281
Geikie, James 123
Genesis, Book of 247, 261, 276
Gentry, Robert 260, 261
Gila River 47, 51, 75, 79, 91
Giotto, space probe 177, 244
Goethe, Wolfgang von 276
Gondwanaland 257
Grand Canal 62, 90, 95, 99
Gravity, gradient of 284
acceleration of 286
twisting of 230, 244
See also Papago Park
Guaymas 36, 40, 44
Guzmán, Nuño de 8, 9
Hallenbeck, Cleve 45, 46
Halos, pleochroic 260 - 262
Hanau Eepe 218, 220
Hanau Momoko 218, 220

Index

Hapgood, Charles 231, 232
Haury, Emil 83, 84, 204
Hearts, Valley of 34, 35
Hemenway expedition 59
Hermosillo 44, 51
Heyerdahl, Thor 222, 227
Higgins, Godfrey xi, 184, 240, 241
Hohokam 73 - 74, 82 - 86
Hokkanen, J. E. I. 319
Holden, Ted 326
Howorth, H. H. 138, 139, 143, 158, 267
Humanism 271 - 275
Hurricanes 169
Hutton, James 149
Hypermatter 270 - 316
Ice ages 119, 120, 251, 255, 257
Indian Bend Wash 99, 102
Irrigation system, ancient 60 - 62, 87 - 93
Isis 183, 184, 188 - 191
James, William 1, 165, 270
Job, Book of 188
Joshua, Book of 176, 185, 192, 238
Jupiter-Baal, temple of 278
Kalasasaya 241, 242, 246
Keilhack, K. 147, 168, 267
Kino, Eusebio Francisco, Father 50 - 54, 58, 70, 72, 73, 193, 208, 211
Laboratory, scientific 2, 4, 6, 148, 163, 167, 271, 276
Lemke, Leslie 268, 269
Light, breadth of 295
Loess 125 - 131, 133 - 135, 138, 140, 144, 146 - 148, 151, 164, 165, 170, 172
   aeolian theory of 136, 146
   humus in 144
   limey slabs on 141 - 142, 265, 266
   nodules in 133 - 134, 154, 156 - 159, 161, 164, 165, 268

   pebbles in 140, 141
   porosity of 130
   snails in 152, 153, 157 - 182
   tubules in 129, 143, 164
   vertical cleavage in 141
Longitude, determining 232
Los Muertos 59, 60, 76, 77, 87
Lugn, A.L. 151, 170
Magnetism, terrestrial 185, 260, 301 - 308
   Rate of decay 303
Mammoths, frozen 317
Mana 218, 221, 280
Manetho 248
Marata, see Cibola, Cities of
Manje, Juan Mateo 54 - 57, 196
Marcos de Niza, Fray 10, 13-33, 35, 36, 39, 41- 48, 51, 63, 65, 70, 72, 74, 79, 85
Marcos de Niza, inscription 63, 64
Maziere 223
McDowell Butte 104, 105, 107, 111, 198 - 203
Mendoza, Antonio de 9, 10, 13, 28, 31
Métraux 221, 279
Midvale, Frank 96
Minerva 188, 189
Moon, distance to 287
   mass of 288
Morrison, David 234
Mount Pelée 313
Mount St. Hellens 315
New Galicia 8, 12, 13, 31, 50
Noachian Flood 247 - 262, 276
Nouée ardente 311
Nuñez, Álvar 9, 10
Oort, J. H. 177 - 181
Other worlds 254, 283 - 286, 292
Overthrust faults 253
Papago Park 91, 96 - 99, 113, 121, 198, 199, 200, 255
   grain size distribution 116

« 335 »

melted rocks 107 - 112
tongue-in-mouth 199, 201
twisted gravity 201- 205, 237 - 238, 244
Park of Four Waters 89, 93, 97
Patrick, H. R. 60
Petrified Forest 253
Petroleum 316
Phaëton 174, 175, 186, 255
Phoenix, City of 59, 62, 63
Pimería Alta 51, 52
Piri Re'is map 231, 232
Plate tectonics 258, 311
Poike Ditch 219, 220
Political correctness 274, 275
Principle of Uniformity
 see Uniformity Principle
Pteranodon 316
Pueblo Grande 62, 90 - 92, 95
Pulsars 300
Quaternary period 253
Quasi-stellar sources 298
Radiocarbon dating 83, 84, 171, 204, 220
Rano Raraku 214, 215, 217 221, 226 - 228, 235
Red-shift 293 - 300
Referees, system of 266
Richthofen, Ferdinand von 129, 130, 133, 136, 138, 146
Roggeveen, Jacob 213, 214, 217
Rokoroko He Tau, King 221, 222, 279
Salt River 60, 61, 78, 79, 91
 excessive erosion in 82 - 85, 118, 210
Santa Fé 65, 67, 69, 70
Saturn 229 - 231, 233
Sauer, Carl 40, 44
Sjkölsvold, Arne 225, 227
Skertchly and Kingsmill 141 265 - 266
Slender People 220, 221, 225
Smith, Carlyle 222

Snails, fall of 168
Snaketown 74, 83, 84, 204
Sodom and Gomorrah 206, 277
Solstice 240, 242, 246
Soul, human 265, 269, 274
South Mountain Park 63
Stonehenge 245, 246
Stout People 218, 219, 222, 223, 281
St. Pierre, Martinique 311
Thunderstorms 169, 173, 190, 259
Tiahuanaco 241, 242, 245, 246, 278
Tides, earth 281 - 292
 theory of 289
Till, glacial 120, 121, 122, 123
Totonteac, see Cibola, Cities of
Transmutation of elements 252, 262
Turney, Omar 80, 81, 87, 88, 89, 102
Uniformity Principle 3 - 6, 149, 150, 249, 256, 261, 263 - 267
Vargas, Don Diego de 66 - 72
Velikovsky, Immanuel 4, 181, 183, 186, 187, 235, 237, 239, 245, 248
Venus 188 - 192, 230
 ashen light 192
 beard of 186, 187, 189, 191
 erratic motion of 187, 190
 horns of 191
 surface temperature of 192
Volcanism 191, 249, 258, 261, 309 - 316
Voyager 2 233, 234
White dwarfs 294, 299
Winchell, N.H. 125, 126, 141, 150, 266
Wurm, K. 183, 184
Year, length of 238, 239
Zuni 39, 42, 49, 67, 77, 78, 80, 85, 315

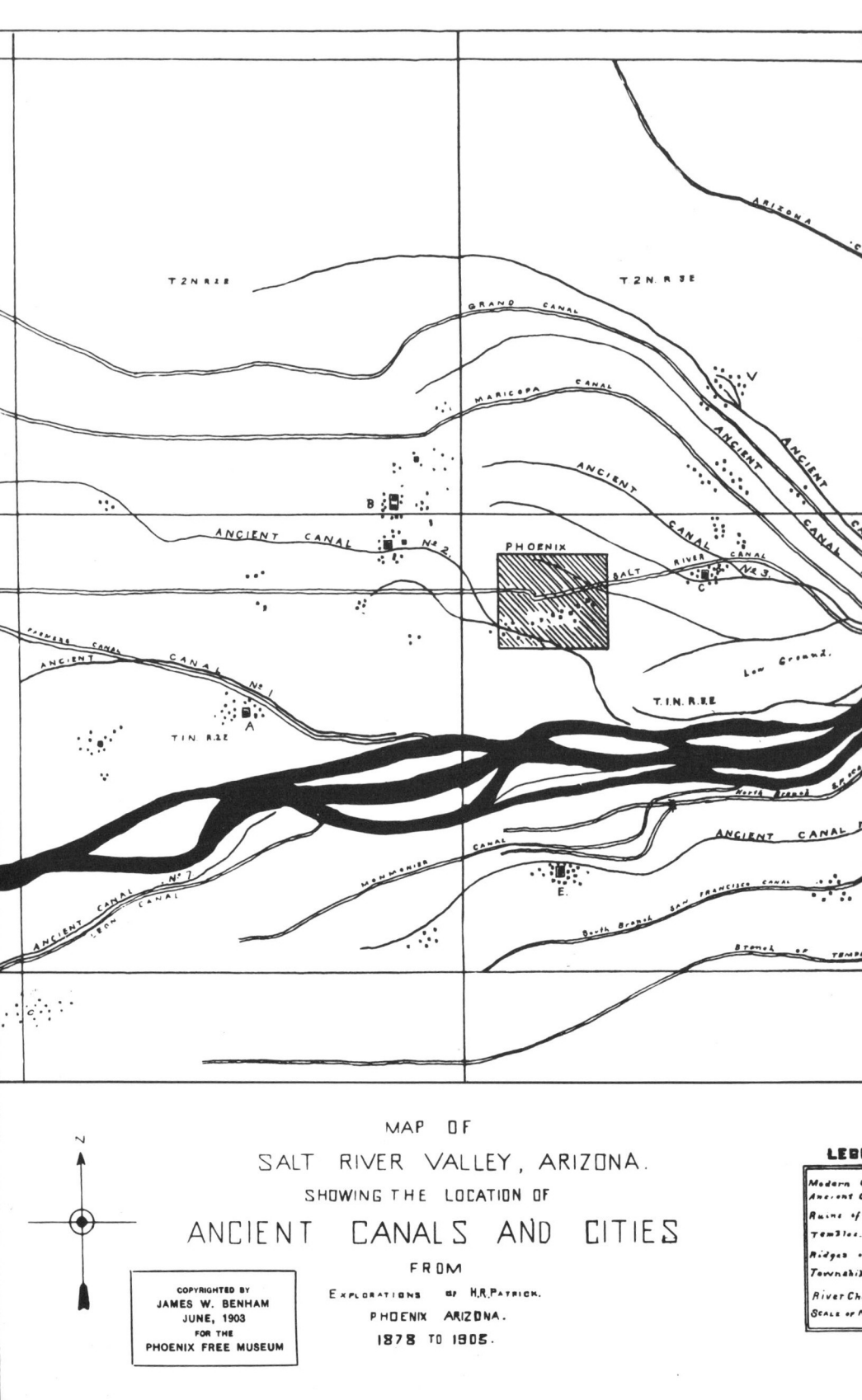

MAP OF
SALT RIVER VALLEY, ARIZONA.
SHOWING THE LOCATION OF
ANCIENT CANALS AND CITIES
FROM
EXPLORATIONS OF H.R.PATRICK.
PHOENIX ARIZONA.
1878 TO 1905.

COPYRIGHTED BY
JAMES W. BENHAM
JUNE, 1903
FOR THE
PHOENIX FREE MUSEUM